A Chronicle
History of

FORTS
AND
FORTRESSES

A Chronicle History of

FORTS AND FORTRESSES

Martin Brice

SILVERDALE BOOKS

A QUANTUM BOOK

This edition published by Silverdale Books,
an imprint of Bookmart Ltd., in 2005

Bookmart Ltd.
Blaby Road
Wigston
Leicester
LE18 4SE

This edition printed 2005

ISBN 1-84509-108-6

QUMFOR

Printed in Singapore by
Star Standard Industries (Pte) Ltd

FOREWORD

Citadel in the Valley of the Tombs, Palmyra, Syria.

Since earliest times, man has been a warrior. Even his first settlements were strategically sited for purposes of defence, while, as knowledge of building skills developed, one of the chief uses to which they were put was in the building of forts and fortifications. This book is a comprehensive record of the development of fortification across the world, from prehistoric times to the present day. Including both technical details and stories of personalities and sieges, *Forts and Fortresses* explains how changes came about, not only in response to new weaponry, tactics and industrial invention, but also through strategic, social, religious and cultural factors.

It is of course impossible in a book of this size and scope to list all the forts and fortresses ever built throughout the world. I have instead chosen the most interesting and historically significant examples of military architecture down the ages. The story that unfolds is illuminated by a wealth of illustrations, maps and diagrams. Some are photographs showing fortifications under contruction, in service and in ruins or undergoing historical preservation of reconstruction. Others are contemporary paintings and drawings, while some are antique representations of earlier fortifications, imaginative rather than accurate, but demonstrating the legendary aspects of castles and strongholds.

Forts and Fortresses is chronological in scope, with each chapter comprising a narrative section and three special features which provide in-depth reviews of particular aspects of fortification in that period. In addition, several box-features highlight individual fortresses, fortified cities or architectural weapons systems of especial significance because of their place in technology, history or legend. The final chapter brings the story up to date and examines the nature of fortification in today's shrinking world. Against a backdrop of futuristic developments for the nuclear age, political perceptions are being reassessed and barriers re-thought. It is indeed an apposite time to reflect upon the history of man's systems of defence and deterrence.

Martin Brice

Martin Brice
March 1990

CONTENTS

Below: (from top to bottom) Angers castle in the disputed territory between France and Brittany was founded in 1232. Vitre, another castle on the borders of France and Brittany. Tour d'Inquisition, City of Carcassonne, southern France. Ypres in Belgian Flanders was fortified in the seventeenth century Vauban period; its ramparts, including the Lille gate, provided shelter to British soldiers in World War I.

Below: (from top to bottom) Castle Barracks, a local headquarters for the British army in Ireland in the nineteenth century. Tripoli – Porta Nuova (New Gate), part of the defences around Italy's stronghold in Libya (photograph taken in the early twentieth century). Castle William, Governor's Island, New York, USA. Inside one of the magazines in Leugenbook a Moere, Belgium, World War 1.

SPECIAL FEATURES

This map shows the location of the special features which are interspersed throughout the book.

Tudor Coastal Defence
p78

Norman Castles
p42

Vauban
p94

American Forts
p114

American Coastal Defence
p134

Gibraltar
p122

Southern Strongholds
p106

Roman *Limites*
p34
Atlantic Wall
p154
Maginot Line
p146
The Great Wall of China
p18
Fortresses of India
p62
Turkish High Tide
p86
Strongholds of the Bible
p22
Crusader Castles
p54

Chapter One

THE ANCIENT WORLD

The earliest fortifications were true earthworks. Deepening an existing gulley and piling the excavated soil along one side is a simple and relatively permanent method of marking a boundary or of making a defensible position more tenable. Even so, such ditches and banks could still be jumped by animal predators and scrambled over by human beings. Indeed, the very passage of such demarcation may transform an approaching figure into a hostile stranger – and into a bitter enemy if the newcomer then makes a mockery of your proud fortifications. According to legend, after all, Romulus was so enraged when Remus leaped over the primitive foundations of Rome that he murdered his brother. It seems an overreaction in the face of apparently childish behaviour.

Such examples, however, demonstrate that more than material desolation is involved. It is the implied or actual threat of destruction to something which symbolizes personal (or national) status: it is a humiliating insult that must be avenged. Failure to react admits submission to the aggressor's will.

Throughout history, therefore, fortresses have been defended in hopeless causes – when to surrender them would mean abandoning the sacred soil of the homeland. Castles have been deliberately erected in disputed territory – a perpetual challenge to the foreigner. Complex examples of military architecture have been

Biskupin (**left**) and Maiden Castle (**right**) represent the fortified communities of the Iron Age, spanning the last five centuries of the pre-Christian era. The aerial view shows the Dorset hillfort's complicated eastern entrance, stormed by the Romans in AD 44 during their conquest of southern Britain. In their heyday, these grassy ramparts would have been revetted with stone and hurdle, and crowned with palisades and watchtowers similar to those reconstructed on the excavated site of Biskupin – a marshland village stronghold some 150 miles (about 240 kilometres) west of Warsaw in Poland. It can be appreciated that such construction (and maintenance) must have demanded considerable social organization.

constructed in impossible locations, to showy design, of flimsy materials, with obsolete armament, by individuals and societies who could not afford them. Simultaneously, there has always been investment in the latest technology to ensure the safety of the fortress and the protection of the builders' way of life. And it is technical advances for which we have most evidence when considering the motives of the earliest fortress-builders.

Both animal and human intruders could be kept out by thorn-bushes, either dead and withered, or by a cultivated quickset hedge. Entrances remained a problem: dragging a spiky tree to and fro is cumbersome and potentially painful – and in any case, attackers can haul it to one side as easily as the defenders.

So the gate was invented, and with it the gate-post and, logically, the fence – surmounting the *inner* rampart. For, in due course, the early builders learned

Left: Lake Van is situated in Turkey, in what was the ancient land of Armenia. In ancient and medieval times this lowland region and its associated valleys and mountain passes was even more important for caravans and armies. Whoever held the strategic points dominating these routes could control trade and deny passage to the enemy, as well as profiting from the farmers labouring in the more fertile soil near the rivers. For over 3,000 years, this natural fortress has served as the Citadel of Van.

to repeat their ditches and banks both horizontally and vertically. They thus presented an attacker with a series of deep trenches and a series of steepening slopes. And if an enemy nevertheless presumed to attack, he could expect all the while to be subjected to a hail of missiles from the concealed defenders.

PREHISTORIC FORTIFICATIONS

The hillforts of prehistoric Europe combined both linear and vertical defence systems, plus a variety of traps, drawbridges, gates and dead-ends. Their banks were smooth slopes of beaten earth, or – if vertically-faced – were revetted with hurdles, roughly-hewn stones, or a combination of both. To a distant observer, the hilltop fortress looked as if it were ringed with walls of solid stone.

Iron Age hillforts in Europe, Maori *pas* in New Zealand, Red Indian encampments in North America, Burmese and Ashanti stockades in the jungle – the earthworks of all or most of these remain, furrowing the countryside or dominating the landscape, although the rulers and communities they protected have long since vanished. Yet in spite of their apparent longevity, these prehistoric strongholds were nothing like as permanent as their builders would have wished. Even in temperate zones, they were indeed *earth*works, returning to the earth from the moment they were completed. Rain-washed soil had to be dug out of the ditches, hurdles and palisades rotted and had to be replaced, weeds had to be uprooted – gently – every season so that bushes and trees had no chance to begin pulling apart the dry-stone reinforcement. Prehistoric hillforts required a programme of non-stop maintenance if they were to remain effective.

Similar repairs were also required on the timber-built fortified lake villages located throughout Europe's marshy regions, and known in Celtic areas as *crannogs*. The problem was even greater in areas of dramatic rainfall and rapid-growth vegetation – the very factors that favoured agricultural fertility, the accumulation of food reserves and the development of material civilization.

These possessions had to be jealously guarded, and the only practical material that offered – and still offers – the fortress-builder reasonably maintenance-free permanence was stone (natural or artificial), shaped and fitted to leave no intervening crevices, and pre-

HATTUSAS

HATTUSAS WAS founded on a hilltop in about 1800 BC near present-day Bogazkoy, some 200 km (125 miles) east of Ankara in Turkey. Its greatest period was when it was the capital of the Hittite Empire in about 1450-1200 BC. It had a stone wall, heavily buttressed or bastioned, in length totalling more than 7 km (4½ miles); its line followed salients and reentrants. There were internal switchlines screening the two citadels, and there were external fortifications covering the approaches to the principal gates. The King's Gate has inward-curving jambs, composed of single pieces of solid rock, which almost meet at the top to form a pointed arch.

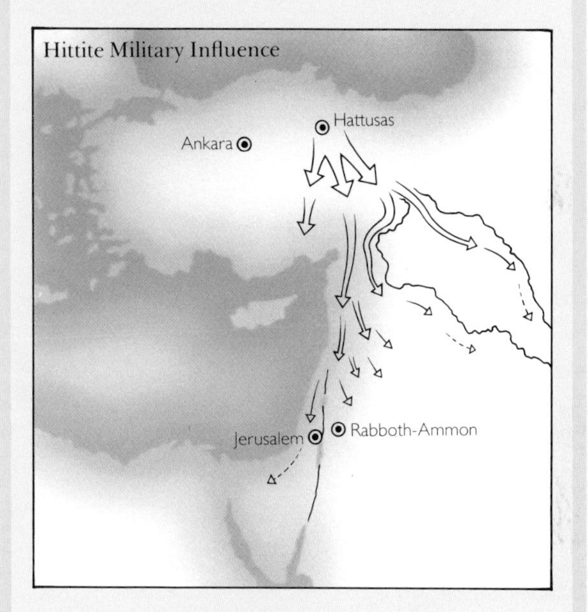

From Hattusas (**below**) Hittite influence spread throughout the Middle East.

ferably bonded and sealed with some pliable substance that itself set as hard as stone. Such a wall will keep out both weather and enemy.

The earliest examples of that sort of solidity are to be found outside Europe, in the Middle East, in India, in South-East Asia, and in China.

None of these civilizations can be said definitively to have been the first to introduce any one particular feature. All were too distantly separated for technical challenge and exchange. Their ancient fortresses were as advanced as contemporary military thought demanded, and local materials and skills dictated. On the whole their construction tended to coincide with the development of a local form of script. For this reason these cultures provide us with the earliest examples of military architecture for which certain archaeological and descriptive evidence remains – whereas reconstructions of European hillforts and lake villages inevitably remains hypothetical.

Such evidence also helps the historian to differentiate between fortified cities as such and those urban complexes whose walls prevented the vulgar populace from defiling the sacred precincts of the god-king. These temple-city walls could certainly be manned in an emergency, but that was not their primary function. After all, the royal divinity was expected to go out and do battle with any rival. If he were a god, how could he lose?

This was the custom on the Babylonian plain around 3500 BC, where the lack of stone and wood meant that brick was the building material most commonly utilized. Each brick was about 10 cm (four in) square and 30 cm (12 in) long. The earliest were dried in the sun, resulting in a whitish colour. Later, redder bricks were baked in kilns, and were much harder, but were also more costly to produce, so that unbaked bricks continued in use. Made of either type of brick, the walls they composed – together with their bonding

Left: The remains of the barracks at Cilurnum (Chesters), a Roman fort about halfway along Hadrian's Wall in northern England.
Right: From AD 70 to 73, Masada, overlooking the Dead Sea in Palestine, was defended by 967 Zealots (men, women and children) against some 15,000 Roman legionaries. Eventually a huge ramp was built, up which an iron-plated siege-tower was pushed, its catapults killing and maiming all who were slow in taking shelter. The garrison still held out, even after a wood-revetted portion of the stone rampart caught fire, weakened and collapsed. The Romans regrouped and stormed in the next day – to find just two women and five children alive; the rest had sacrificed themselves rather than fall into the hands of their sacrilegious oppressor. This view of the Snake Path approach gives a good indication of Masada's precipitous cliffs.

of clay, lime mortar or bitumen – had to be very broad-based, as thick as 15 m (50 ft), to support the weight of as many as 500 rows of bricks above. The result was that they could not easily be undermined. The shortage of suitable timber meanwhile prevented the construction of scaling-ladders tall enough to reach the top of the wall and simultaneously strong enough to bear the weight of several armed men all at one time. (To ascend one at a time would be to invite a succession of pointless deaths as each man reached the top.)

Farther north-west, in mountainous Mesopotamia, Asia Minor and Canaan (Palestine), building stone was plentiful and forests locally accessible, but fertile land was scarce and not sufficient to support populations which multiplied in fat years and which competed for food in lean years. In these territories war was not waged to secure the death of the opposing ruler and the voluntary or compulsory acquisition of his subjects to

the greater glory of the victor; here slaves, no matter how skilful or hard-labouring, were simply extra mouths. Here the object of warfare was the slaughter of aliens and the seizure of their crops.

Accordingly, hilltops were crowned with clusters of stone huts, their interlocking walls presenting a forbidding face to the foreigner. Settlements in which tribal chiefs resided increased in size to accommodate his priests and warriors, his craftsmen, his foodstocks, and the refugees who hurried thither when danger threatened. These complexes were surrounded by purpose-built walls with ramparts and parapets, from which stones, fire-hot sand and flaming lumps of tallow and bitumen could be dropped upon the hostile.

Compared with stones, arrows were long-range and accurate missiles. It was the need for bowmen to take aim that resulted in the development of the classic notched style of parapets: crenellation. Projecting towers and lower-level bastions covered the dead ground at the very base of the wall, while complicated gateways bewildered and doomed the attackers if their battering-rams, axes and picks did succeed in breaking through the first barrier.

The very last resort was for every man, woman and child to crowd into the very strongest tower and to barricade the door. However, the words 'They utterly destroyed all that was in the city, both man and woman, young and old, and ox, and sheep, and ass, with the edge of the sword. And they burnt the city with fire, and all that was therein' occur too often and too specifically in the Old Testament and other ancient writings for them to be dismissed as mere hyperbole.

The enclosed mountain settlements were the Middle East's equivalent of European hillforts. They represent the decidedly defensive concept of contemporary military architecture. Individuals, tribes or whole nations

Left: The gate has been demolished by modern weaponry, but the ancient walls of Nanking still stand after the Japanese capture of the Chinese capital: 13 December 1937.
Right: Masada, showing the Roman siege-ramp.

retreated with their real or spiritual treasure inside a fortress complex, holding out and gathering strength until ready to unleash a devastating and victorious counter-offensive against a foe exhausted by continual battering against impregnable defences. In peace, each one of these fortresses was a self-contained manufacturing and trading focus for the surrounding rural community. It was only when directly besieged that the city and its walls became of martial significance.

The ancient Egyptians, however, developed architecture of purely military function: a system of tactical forts for strategic purposes. It is true that some of their cities had walls – but these were more to enhance and preserve the sacred prestige of the Pharaoh than to ward off foreign besiegers.

Any invaders, after all, had a long way to come. From the north they had to cross the Mediterranean Sea and risk being intercepted by Egyptian warships. From the south they had to march or sail down a long, winding, cataract-obstructed river valley. Here the bowmen stationed on the walls of a few skilfully-sited forts dominated both the Nile's gorges and landing-places, and the south-north highway running along the bank. These strongpoints could be bypassed only via a lengthy detour through the inhospitable and trackless wilderness. Nor could the invading host rush past, accepting casualties but otherwise ignoring the fort. They had no idea how many troops were accommodated inside ready to emerge and harry the intruders into an ambush by Pharaoh's field army – or ready to cut off a subsequent homeward retreat.

No, the invaders would first have to capture the forts, most of which (like Semneh) were visibly decorated with such challenges as 'This is my fortress here. No black man shall pass north of it. I am King and what I say, I do.' A Nubian leader seeking to liberate his people from Egyptian rule could hardly ignore such a political and racial insult. So the attackers swarmed across the dry ditch to assault the walls and gate with picks and crowbars – only to find themselves at the mercy of archers firing vertically down through the balconies which topped the walls.

Invaders from east and west likewise had to traverse many leagues of barren desert. Although small raiding parties might get through, great armies could not do so without organizing the logistics necessary to ensure survival as a fighting force. And long before that could be arranged, Pharaoh's intelligence service would have learned of the plans and the Egyptian army be prepared for battle.

The Egyptians were not always victorious in battle, but when defeated they were ready to learn from their conquerors. By the beginning of the New Kingdom in 1567 BC (which succeeded the occupation by the Hyksos), the Egyptian army had adopted two new weapons systems from their temporary rulers. The double-convex composite bow of wood, horn and sinew was so bound or glued together that at rest it seemed to be bent the wrong way round – but when strung and pulled it had a range of 400 paces. Simultaneously came the single-axled chariot that had spoked wheels, a high-speed and manoeuvrable platform for archers and spearmen.

To accommodate these fighting vehicles and the new rapid-fire infantry, small forts (later known as *migdol*) were built at oases and road junctions. Wide-ranging patrols covered the desert, while the garrison archers prevented hostile movement along the road to the Nile. If a large army approached, charioteers were despatched, via relays of staging-posts and chariot-

THE GREAT WALL OF CHINA

The largest fortification in the world resisted the onslaughts of the Mongol hordes for hundreds of years

The earliest great barriers around Chinese territory were constructed during the Period of the Warring States which began in 476 BC. Several principalities threw up ramparts of earth and stones as boundary markers and as military obstacles against incursion.

One system ran from the Gulf of Pohai near Yungkow northeast to contain the Liao floodplain, and then westwards to beyond Paotow in Suiyan. The total distance was about 1,800 km (1,100 miles.) Another system covered the approaches to Qin from the Wei to the Yellow Rivers, a switchline paralleling the bank of the Luo River.

Some 200 years later, under the dictatorship of the great Qin Shi Huangdi, the peasants were conscripted to build a wall. It mainly followed the route of the north-

Above: Ceremonial banners wave, where Imperial troops once signalled with flags, smoke and fires.

Below left: Earthworks and other barriers during the Period of the Warring States, 476-221 BC.

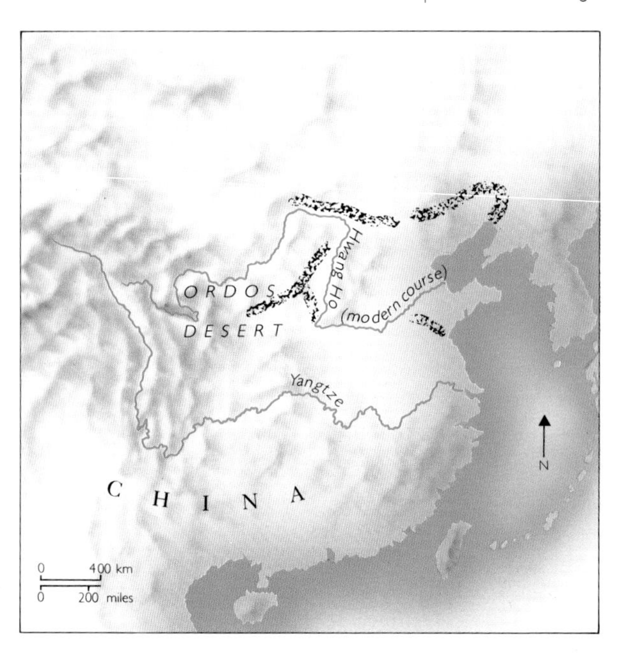

The Great Wall of Qin Shi Huangdi, 221-206 BC.

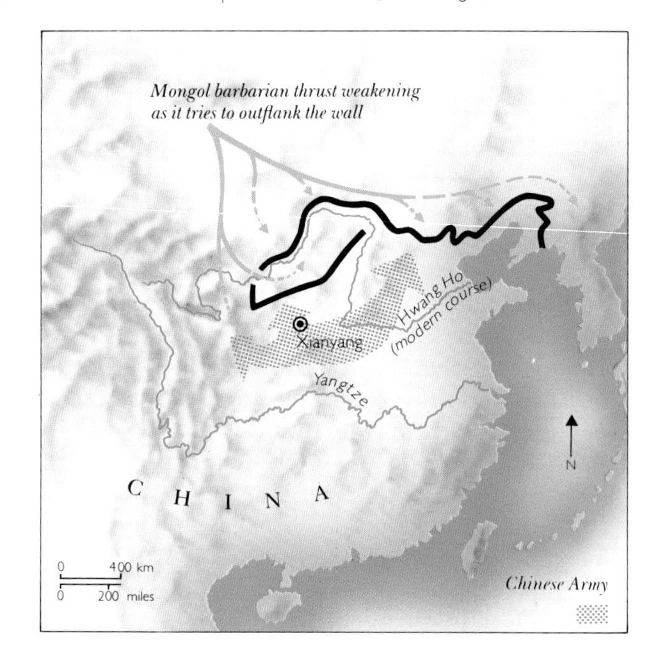

ernmost ramparts from the Gulf of Pohai to Paotow, but then veered southwest to embrace the Ordos Desert and terminate near Lanchow (Lanzhou), a total distance of 4,000 *li* 3,200 km (2,000 miles). Basically an earth bank, although its squared turves probably gave it a similar appearance to any masonry sections, it must have been turreted at intervals with wooden, stone or earth parapet. It must also have had gateways.

Those occasions thereafter when foreigners broke into China seem to have been the result of treachery at a gateway, the defeat of the imperial field army at a place of its own choosing, or the breakdown of Chinese government, rather than the failure of the Wall as a defence system. Dilapidation seems to have been due not so much to enemy action but to earthquake, erosion or neglect.

The Han Emperor Wu-Ti (140-86 BC) ordered a 900-*li* extension running northwest from Lanchow to Jiayuguan. It hindered hit-and-run raids by desert bandits on the western road which carried both the silk trade and military traffic supporting the Chinese advance into Sinkiang.

The Han Dynasty ended in AD 220. Their successors tried to maintain frontier defence, but the Great Wall fell into disuse and was abandoned altogether in 1004.

The Mongols were driven from Peking in 1368 and the Ming Dynasty was founded. It is to this period that we owe much of the Great Wall's present appearance. Some sections were allowed to decay completely, but most of it was rebuilt in stone. Wherever possible, the earth rampart was tidied and given a masonry cladding.

The Wall thus has a base that measures almost 10 m (33 ft) at its widest, and a height that varies from 7.5-12 m (25-40 ft), with a crenellated parapet rising a further 1.5 m (5 ft) above the dressed stone and brick wall proper. Towers rising 3.5 m (12 ft) above the wallwalk are located about every 200 paces, close enough for crossbowmen to cover the intervening stretches of wall – on both the Chinese and barbarian faces, for the towers project on both sides. The stone-paved top of the wall is 4 m (13 ft) wide, allegedly capable of permitting cavalry regiments to gallop five abreast.

The old earth rampart built by the Hans out to Yumen was allowed to decay, the Ming Emperors starting the refurbishment of the Great Wall from Jiayuguan in Kangsu Province. From Jiayuguan the Wall followed the previous fortification southeast towards Lanchow, then northeast to Yinchwan, and zigzagged southeast to meet the line of the oldest Qin frontier boundary near Tingpien. The Ming Wall followed this all the way to cross the Yellow River and pass the garrison city of Tatung (Datong) before arriving north of Peking. The stretch from here to the sea near Chingwangtao (Qinhuangdo) was a new construction by the Mings, resulting in a total length of 2,600 km (1,600 miles).

The Ming Emperors maintained and garrisoned the Great Wall until 1644. In that year, a bandit named Li Tzu-cheng seized Peking, whereupon the last Ming Emperor (Ssu Tsung) committed suicide in shame. Still loyal to his dead Emperor, but unable to defeat the usurper, General Wu San-kuei appealed to Dorgun, ruler of Manchuria, outside China proper. And so the Great Gate of Shanhaiguan was opened to admit the Manchu Dynasty and Wan Li Chang Cheng – 'The Long Wall of Ten Thousand *Li*' – ceased to have military significance.

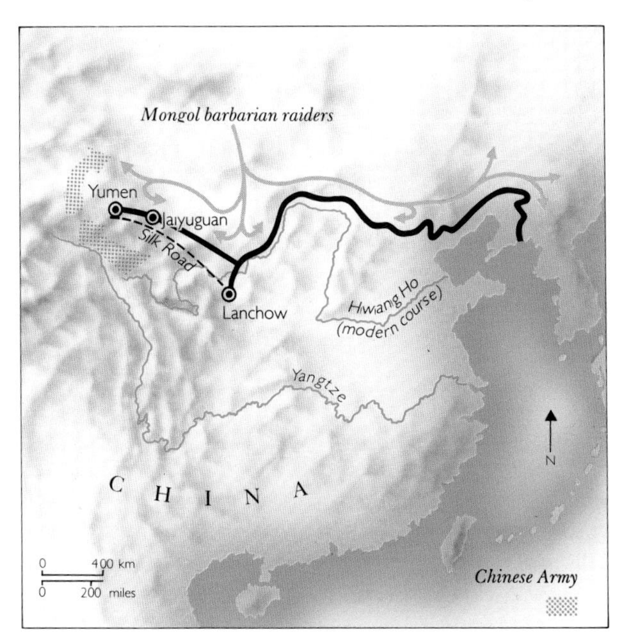

The Great Wall of Wu-Ti (140-86 BC) and other Han emperors.

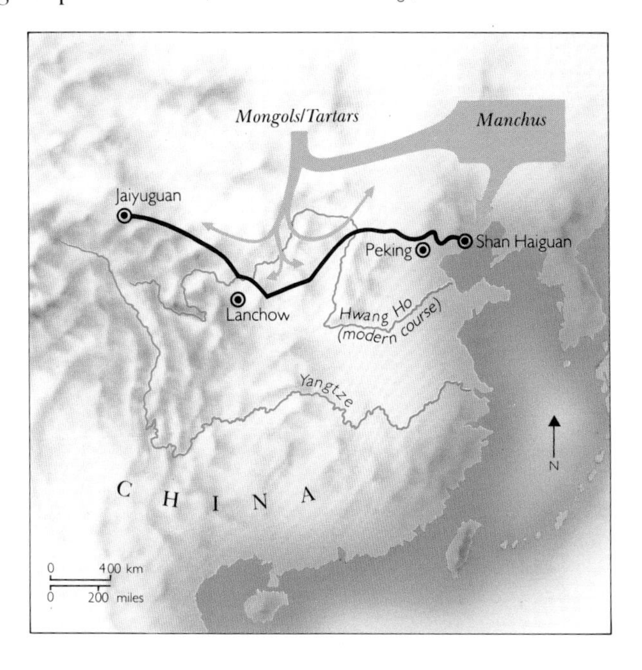

The Great Wall of the Mings, 1368 AD-1644.

MYCENAE

Each City-state in ancient Greece had its own mountain stronghold or *acropolis*, its natural defences improved with stone ramparts. The palace-city of Mycenae (between Corinth and Argos) was founded in about 3000 BC, attaining its full splendour around 1350 BC. Crowning a precipitous cliff, the citadel comprised houses of stone and wood for the royal entourage. The palace itself was a complex of state chambers, public rooms and storehouses, plus a sacred enclosure and tombs, all located on the highest point. On terraces around the acropolis were gardens, vineyards and orchards. The whole self-contained community was surrounded by a wall of massive blocks known to later Greeks as 'cyclopean stones' – suggesting that the creator had been the Giant Cyclops. The ceremonial entrance to Mycenae was the famous Lion Gate, 3 m (10 ft) high and 3 m (10 ft) wide.

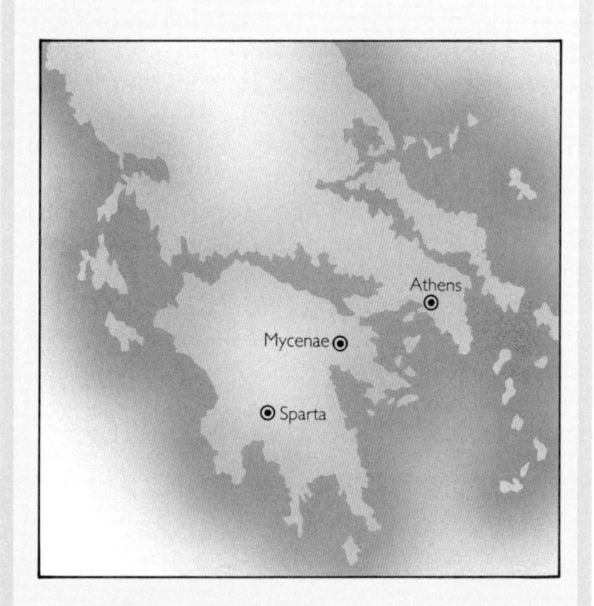

The map **above** shows strategically situated Mycenae (see **below**), Sparta (isolated in the middle of the Pelopponisos) and seaward-looking Athens.

repair workshops, to alert Pharaoh's field army.

These *migdol* and the Upper Nile defences were not refuges for frightened peasants led by boastful warrior-heroes, nor fortified centres of trade and industry. They were garrison communities manned by full-time soldiers, without whom these buildings were worthless shells devoid of military significance. They represent the offensive employment of military architecture – the use of forts to prevent or delay the enemy's movements and to force him into a disadvantageous position where he can be struck by your own field army, itself often operating from fortified strongholds.

Egyptians, Hittites, Assyrians and Persians – these empires successively dominated the Fertile Crescent, eliminating lesser peoples, defeating each other in pitched battle, and in their own ultimate defeat hiding in their last remaining fortress, awaiting the final assault of the vengeful besiegers. To postpone that inevitable day, walls increased in height and girth, their line a convolution of salient and re-entrant so that no scrap of ground was safe from flanking fire. They were fronted with a broad slope or glacis of naked rock or bare earth. Entrance gates were enclosed in their own fortresses. Water supplies were guaranteed; food stocks hoarded. A massive citadel afforded further security for the king and his entourage. Even within that complex, a secret room inside the well-defended turret in a strong tower protected the king alone.

But it was all to no avail. New siege equipment and fresh techniques, operated by massed armies of full-time soldiers (usually conscripts, but no less disciplined and efficient for that), directed by ambitious leaders in a policy of calculated ruthlessness, meant that no city could withstand a siege for ever.

Up to this time, all these developments had occurred in response to the activities of immediate neighbours. Alexander the Great changed all that. Although his empire died with him in 323 BC, by then his Greeks had travelled the known world, reaching the borders of India, bringing their questioning intellects to bear upon what they observed and learned. That argumentative spirit did not end with Alexander's death, nor was military architecture excepted from research and debate. A school devoted to its study was founded on the island of Rhodes by 200 BC. Knowledge was written down and preserved, diagrams prepared, defences discussed, siege-engines designed, and theories tested.

THE TRIUMPH OF ROME

Among those who studied there and made use of its scholars were men from Rome. They incorporated what they learned in the fortification of their own cities – even in peaceable, settled areas of the empire. Town walls and massive gates were symbols of civic status and imperial pride. The Romans also used what they learned to improve their own siege-engines and techniques – and ignored the more fanciful projects of pure theorists. Above all, they made sure that equipment and personnel (artillery, engineers and transport) were properly organized.

These troops came under the *praefectus fabri* ('Chief of Artisans'), who was responsible to the *Imperator* ('General'). He then allocated personnel and equipment to the legions in his army so that any one legion commander (or *legatus*) could call upon the fire-power of some 60 *catapultae* of various types and sizes.

When besieging an enemy stronghold, the Romans first dug a *circumvallatio* all the way round it to prevent the defenders from escaping. A basic *circumvallatio* comprised a simple ditch and bank, but more complex ones had several entrenchments, palisade-crowned ramparts, redoubts and stakes. Similar, but farther out, and so dug that it presented its obstacles in the opposite direction, was a *contravallatio*, erected to protect the Romans against attack from outside by the enemy's field army.

Inwards from the *circumvallatio* was erected an *agger*, a ramp of earth and hurdles, reinforced with tree-trunks and fireproofed with dry-stone facing. It was designed to reach to the enemy's ramparts or walls. Up it was trundled a wheeled tower, up to ten storeys high (during the medieval centuries known as a 'belfry'). The bottom level carried a battering ram – of which there were two types: one to hammer at an obstacle, the second to lever like a giant crowbar. Other floors in the siege-tower were equipped with hooks and iron-pointed levers to rip or prise away masonry and palisades.

Grapnels could also be used to hurl and affix rope-ladders, in addition to the more common rigid scaling-ladders. The largest siege-towers had drawbridges that could be dropped on to the enemy's battlements.

The topmost level mounted a catapult weapon. Anti-personnel *scorpiones* were big crossbows that fired

TROY

Ulysses (right) carrying the armour of Achilles.

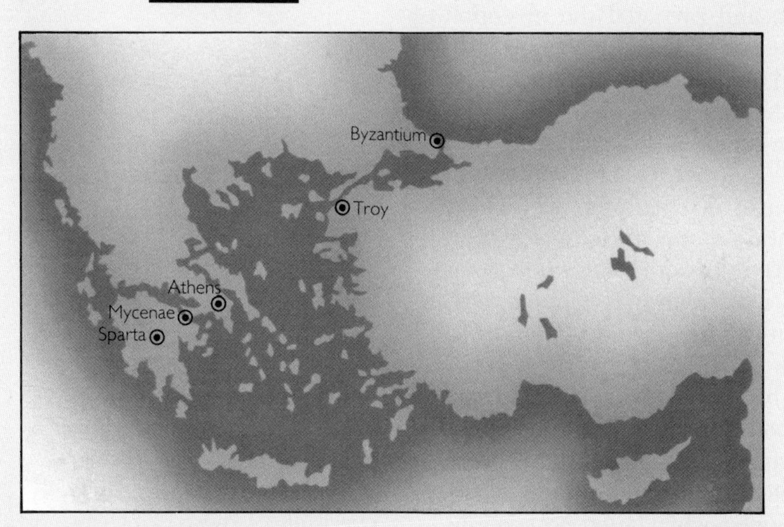

There are at least nine consecutive cities, the earliest dating from 3000 BC, under the mound at Hissarlik in Asia Minor overlooking the Dardanelles.

The sixth city had very steep-sloped stone walls, which are still 6 m (20 ft) high. There were several towers. One of the principal gates was formed by overlapping the plan of the wall so that the entrance was invisible to an attacker.

Either this or the seventh city is believed to be the site of the legendary siege of Troy (or Ilium), said to have occurred in about 1184 BC, and to have been won by the Greeks only after the Trojans dragged the Wooden Horse containing Ulysses and his commandos into their city. Troy was destroyed and its inhabitants massacred. Only Aeneas and a small band of Trojans survived to wander westwards, found the Roman Empire, and one day conquer Greece.

The story of Troy illustrates not only the single-combat method of conducting sieges in Homeric Greece, it shows also how one period's legends influence later reality. This was a very popular story in medieval times and helped to shape chivalric ideas about sieges, heroism and correct behaviour. It also has a twofold significance for the study of warfare generally, even today. No matter how impregnable the fortress, it can always be taken by a clever trick; and no matter how extensive the massacre, somebody always survives to tell the tale, keep the memory alive, and one day return in vengeance.

STRONGHOLDS OF THE BIBLE

Then as now the Near East was rent with conflict. The walls of Jericho were a vital defence against neighbouring enemies

Excavation of the remains of the ancient city of Jericho (located on the west bank of the valley of the River Jordan) has revealed at least 19 layers of urban construction, many of the walls showing signs of demolition by earthquake or war. The earliest settlement dates from the Neolithic period, about 7000 BC. It covered 4 hectares (about 10 acres) and predated the invention of pottery; its 3,000 inhabitants fashioned utensils out of limestone. And yet they enclosed their town with a free-standing stone wall 2 m (6 ft) thick, 7 m (23 ft) high, and 800 m (½ mile) in circumference. There was also a citadel-refuge tower 7.5 m (25 ft) in breadth. The whole complex was surrounded by a ditch 8.2 m (27 ft) wide and 2.7 m (9 ft) deep.

This layout became the usual method of fortifying a city in ancient Palestine. Of course there were variations as weaponry and tactics developed, when larger populations required additional reserves of food and water, or when the city was intended also to serve as a royal base for military operations against neighbouring states.

When King Solomon (ruled 974-937 BC) introduced horses in place of mules to draw war-chariots, he chose Megiddo on the slopes of Mount Gilboa as his northern base of mobile operations. The stables there (probably rebuilt in the slightly later Omri-Ahab period of 885-853 BC) comprised five sheds with two rows of 11 stalls in each. The roof was supported by stone pillars,

Megiddo and Trade Routes

Haifa
Sea of Galilee
Tell Keimun
PLAIN OF ESDRAELON
Megiddo
VALLEY OF JEZREEL
Jordan
Beth-shean
WADI ARA PASS
0 10 20 km

The map **above** shows how hilltop Megiddo was well-placed for its mobile forces to control the valleys radiating from the Plain of Esdraelon.

Right: A more general map of Palestine.

which also served as hitching-posts. In front of the compound was a courtyard with a large cistern for watering the horses.

The walls of these cities were of sunbaked brick or of stone, cut and fitted together with or without clay mortar. The outside of the wall was often covered with lime plaster, making it more weatherproof, more difficult to scale, neater, and more impressive. The whole complex was surrounded by a glacis which became one of the most important defensive features: smoothing the natural incline or piling up pounded earth at the base of the walls produced an exposed killing-ground too steep for the safe footing of scaling ladders. Further obstacle was provided by the 20-m (65-ft)

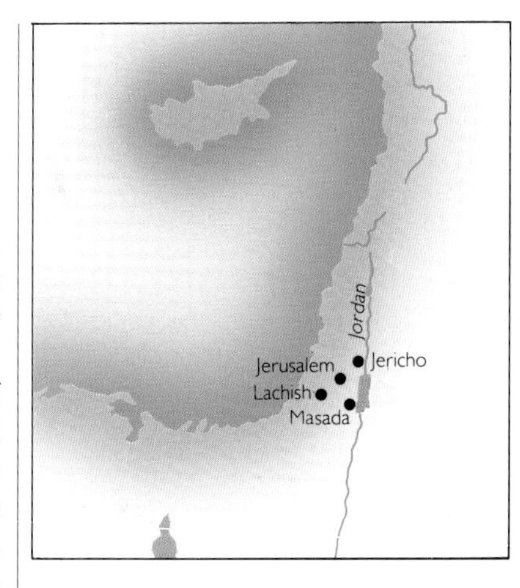

vertical shield-wall of cyclopean masonry which held the glacis in place.

At Lachish (near Gaza in Judah), the glacis shield-wall was heightened, probably in about 926 BC. It now formed a second free-standing full-scale wall, complete with crenellation, projecting towers and its own fronting glacis and retaining wall. By 704, the walls had been crowned

Masada: the Round Pavilion on the Middle Terrace of Herod's Hanging Palace. The double-columned Lower Terrace, affording panoramic views out across the Dead Sea, is towards the top of the picture.

with a wooden balcony from which the archers could fire vertically downwards upon the assaulting troops.

With this concentric layout, the gatehouse defences became even more complex, for they now provided the only practical means of reinforcing the defenders on the outer wall or of withdrawing them if that became necessary. The sole means of approach was up an inclined ramp, which ran parallel to – and exposed the attackers' right unshielded side to arrows from – the two overlooking walls. Battering into the fortress-gateway, and turning to their right, the besiegers now found themselves looking at three more doors, each composed of solid wooden boards strengthened with metal plates and secured (from within) by wooden beams and bronze or iron bolts. Two led on to the glacis on either side of the gatehouse and between the two walls. The middle gate was the true entrance to the city, and also the most logical, but it was not always possible to think clearly while the defenders – hidden in secret corridors and rooms in the hollow walls and ceiling – poured down a deadly rain upon the mass of men in the semi-darkness between the gates.

But the strongest fortification is of no account if the defenders die of thirst. To guarantee Jerusalem's water supply, King Hezekiah of Judah (ruled 719-691 BC) organized the excavation of an underground aqueduct from outside the city. When the Assyrians arrived, they called upon the citizens to overthrow Hezekiah and come out. No response forthcoming, the Assyrians settled down for a long siege...and eventually had to abandon it, after being struck by disease.

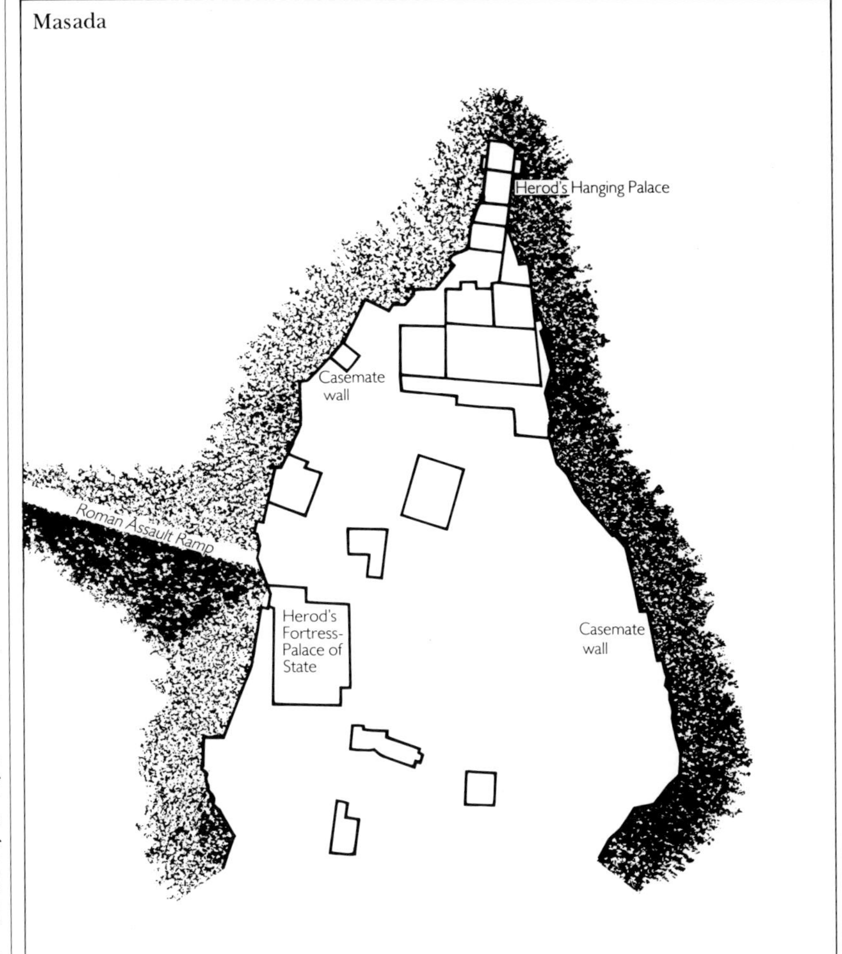

The first stronghold at Masada was a fort built by Alexander Jannaeus (103-76 BC). In 30 BC, Herod the Great chose the site for the construction of a luxurious refuge. Access was via a track (later buried under the Roman siege-ramp) to the east. This approach was dominated by Herod's fortress-palace. Right at the northern tip of the plateau was Herod's Hanging Palace, while other buildings served for stores and accommodation. The whole complex was surrounded by a casemated wall. Modern access is via the Snake Path (or cable-car) from the west.

metal bolts up to 3.5 kgs (7½ lbs) horizontally or at a slight angle. The higher-trajectory *ballistae* hurled stones and lumps of timber, mainly for material destruction but with anti-personnel effect when the missiles shattered on impact. Propulsion was provided by a wooden arm embedded in a mass of twisted sinew or hair – women's hair was found to be the longest and most resilient. The arm was wound back and a 25-kg (55-lb) stone placed in a sling at the end. The arm was released, flew forward and hit a massive cross-timber like a goalpost, whereupon one end of the sling flew free and the missile shot through the air over a distance of 200-400 metres (220-440 yd). The chief disadvantage of the *ballista* or the *scorpio* was that the hair propellant could be adversely affected by rain.

SPARTA

CONTRARY TO what might be expected, Sparta – the most militaristic city-state of ancient Greece – had no fortifications. According to tradition, Lycurgus (the legendary compiler of the legal constitution of classical Sparta in about 776 BC) forbade the construction of defensive walls, thus forcing the citizens to trust in their own valour for their protection. Sparta's contribution to military architecture was the creation of the barrack block, in which all male Spartans between the ages of seven and 30 lived under martial discipline.

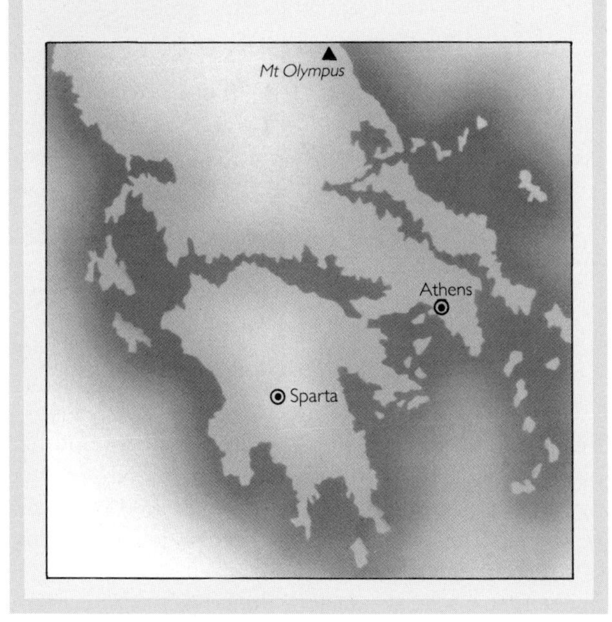

Above: Map showing the location of Sparta.
Left: This photograph of a reconstructed *scorpio* was taken at a display of the Ermine Street Guard at Fishbourne Roman Palace near Chichester (Noviomagus) in southern England.

All these weapons could be employed separately while protection was provided by *testudines* ('tortoises'). The simplest *testudo* was formed by legionaries' locking their shields over their heads and around their bodies. More substantial ones were built for prolonged siegework. The *musculus* was a solid penthouse roof that was leaned against an enemy wall to screen engineers as they tunnelled under the masonry.

These devices had been known for several centuries before the Romans, and were to continue in use until the introduction of gunpowder and cannon.

There was also the weapon referred to by later generations as 'Greek Fire'. It seems to have first become famous in about 415 BC, when it was employed by the Greeks of Syracuse in defence of their Sicilian colony against the Athenians. There are two theories on what it was made of. One is that it was an incendiary mixture of pitch, sulphur, tow, frankincense and pine-sawdust, placed in pots in *ballistae* slings, ignited and hurled towards the target. It could also be wrapped around the heads of fire-arrows. However, there are also suggestions that it was emitted by some sort of tube for projecting flame, and that its principal constituent was naphtha or some other flammable chemical derived from fossil oils that occur naturally in the Middle East.

Whether the secret of its manufacture was subsequently lost, or whether it was so terrible a weapon that the use of liquid fire was banned by tacit agreement is not clear. What is certain is that the first definite references to its existence date from AD 670. And it is immaterial whether Kallinikos (a Greek living in Heliopolis in Syria) learned its secrets from some long-lost manuscript or whether he actually created a new formula and means of delivery. What he did was to invent a bellows or pre-pressurized container which, when triggered, projected the deadly concoction like a modern-day flamethrower. A citizen of the Byzantine Empire, Kallinikos made his discovery available to the armed forces. A refinement was its installation in the fireproofed bows of ships for maritime operations. A much smaller projector was also made, and could be hand-held for siege operations in both offence and defence.

The manufacture of 'Greek Fire' could be problematical, and it seems not always to have been available to the Byzantine forces. It may have been more hazardous (or at least more temperamental) than gunpowder – it was said to continue burning even when water was poured on it. Certainly its secret died with the fall of Constantinople in the fifteenth century AD.

THE SIEGE OF JERUSALEM

Such devices (but not Greek Fire) were employed by the Romans during their siege of Jerusalem in AD 70. They had hoped that the area might have been a comparatively trouble-free part of their empire – after all, they had given support to the Maccabees during their successful rebellion against the religious and political domination of Hellenistic Syria. Moreover, from 40 BC onwards, the Roman protegé Herod the Great had ruled Palestine. He occupied Jerusalem as his capital in 37 BC, and raised and strengthened the walls there and built two fortresses.

One, surmounting a 25-m (80-ft) crag at the northwest corner of the splendidly rebuilt Temple, was named Antonia in honour of Herod's friend and patron, Mark Antony. It was uncompromisingly square. It had four square corner towers, overlooking the Temple itself, and accommodated the Roman Occupation Force, and enabled them to watch the likeliest trouble-spots in the city.

The other citadel also served as Herod's palace, and was both opulent and fortified. It had three huge square towers named Mariamne (after Herod's wife), Hippicus (his friend), and Phasael (Herod's brother). Each was a solid mass of cut boulders 3 m (10 ft) in dimension, to a height of 20 m (70 ft), with living accommodation to a similar altitude above that. Phasael was the biggest, looking, so it is said, like the Pharos of Alexandria.

By this time Jerusalem had two walls: the First or Broad Wall encircled the Upper City (which included the original city of David), crowned the steep ascent to the south, and, on the north, ran straight east-west from the Temple and the Antonia to Herod's palace. From the Antonia, the Second Wall curved in a northerly arc to enclose the Lower City before also reaching Herod's palace.

By AD 41 'Hierosolyma' (as the urban complex was now known) had spread farther north. So Herod Agrippa I decided to build the Third Wall to protect the New City. However, such an undertaking seemed a hostile act to his suspicious Roman patron, the Emperor Claudius, and work stopped.

Herod Agrippa II (who succeeded in AD 44) was very much a creature of the Romans, even as his Jewish subjects (fired by the Zealots) grew increasingly resentful of the betrayal of their nationhood and religion. In AD 66 they overwhelmed and massacred the Roman garrison in Jerusalem. The rebellion spread through Judaea and the Roman legionaries suffered serious defeat. Meanwhile, work had been resumed on the Third Wall of Hierosolyma. When complete it was 4.5 m (15 ft) wide and 10 m (33 ft) high, its square rain-collecting and fortified towers rising another 10 m (33 ft) above that. The biggest tower was octagonal, reached 33.5 m (110 ft), and was called Psephinus. By the spring of AD 70, 340 catapults had been mounted on the walls. Led by John and Simon, and augmented by pilgrims arriving for the Passover, 30,000 Jews were ready for the onslaught of three legions (totalling 60,000 soldiers) under Titus.

Right: The ancient seven-gated wall surrounding the Old City of Al-Kuds (or 'The Holy One'), as Jerusalem was known to the Ottoman Turks, was rebuilt by Suleiman the Magnificent after 1520. By then the wall had lost most of its belligerent purpose and was more useful for municipal administration and trade regulation. Nevertheless, some military features can be seen in this view of the Damascus Gate. The paired pointed merlons with arrow-slits between were fashionable throughout the Muslim world. Other arrow-slits have extra round holes for early hand-guns, while older-style machicolation has been built out on decorated corbels. Though vulnerable to twentieth-century artillery, such fortification still provides useful protection against anti-personnel weapons, as was learned during the operations in Jerusalem following the establishment of the state of Israel in 1948.

At first it was an artillery duel, each side trying to wreck the other's heavy catapults, culminating in a Jewish commando raid which both burned the Roman *ballistae* and the mantlets protecting the battering rams, and killed the engineers and sappers.

The Romans then constructed three iron-clad wooden siege-towers 10 m (33 ft) high, and mounted lighter catapults on top. Pushed into position, one tower collapsed under its own weight (not through sabotage, as the Romans at first suspected). Because the southern approach to the walls was hindered by

precipitous slopes, the Romans concentrated against the northern Third Wall. The two surviving siege-towers provided outranging covering fire for the biggest *ballista* ('The Conqueror'). This finally smashed through the Third Wall, whereupon the Jews retreated through the New City to the Second Wall. At this stage the siege had lasted for 15 days.

Now the Roman artillery and battering ram hammered away at a tower in the middle of the Second Wall. A distraction was caused by an altercation on the wall between factions for and against surrender. Titus halted operations pending a conference, but missiles were hurled at the Romans as they approached, whereupon the legionaries resumed their battering. The tower started to collapse. The Jews set its wreckage on fire, and retreated through the alleyways of the Lower City, fighting rearguard actions which sometimes turned into counter-attacks. Eventually they were forced back within the First Wall of the Upper City. The siege had now lasted 20 days.

Four Roman siege-towers (which suggests good stocks of timber in Palestine at that time) then battered the walls. The Jews retaliated in similar fashion, and dug a tunnel out from under the eastern defences. When they calculated that they were under a siege-tower, they set fire to the props supporting the tunnel roof. It collapsed, bringing down the siege-tower above, whereupon the Jews on the western sector sallied forth, rushed the other Roman equipments, and burned them too.

Before his next assault, Titus had his troops construct a *circumvallatio* all round the city, including the

BABYLON

A SETTLED urban complex by 4000 BC and capital of many empires centred on the floodplain between the Tigris and the Euphrates, Babylon achieved its greatest fame under the Chaldean Nebuchadnezzar II (ruled 604-562 BC). Its surrounding wall was 18 km (11¼ miles) in length, with a reinforcing tower-buttress every 20 m (22 yd). The inner surface of the wall was composed of 6 m (20 ft) of unbaked brick. Then came 9.2 m (30 ft) of rubble, retained by a fronting wall of baked brick 7.6 m (11 ft) thick. The surrounding moat was fed by the Euphrates, which flowed right through the centre of the city. However, Nebuchadnezzar had learned the lesson of the Assyrian capital – Nineveh – where the foundations had been washed away by flood; the urban side of Babylon's moat was lined with 3 m (10 ft) of brick. The wall's greatest weakness was its number of gates, eight in total, and of splendid magnificence.

The city's opulence (which included the wondrous Hanging Gardens) meant that the name of Babylon later became a byword for any great – and sinful – metropolis.

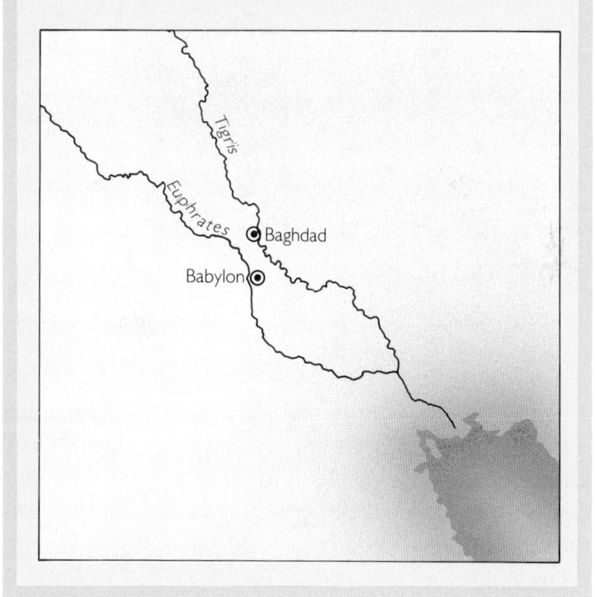

Above: Map showing the location of Babylon.

Left: A pictorial reconstruction – dramatic rather than accurate! The Roman battering-ram is too ornate; the machicolation lacks murder-holes; such external buttresses would be vulnerable; and what is the enemy doing about it?

Siege and counter-siege in ancient and medieval times. The besiegers have erected a ramp of logs, earth and cut timber. On it they have constructed a siege-tower. It is too far away to reach the walls, so the ditch will have to be filled in before it can be pushed near enough for any sort of assault bridge to be used. Meanwhile the main army is drawn up ready to rush the breach or intercept a sortie by the garrison. The latter are replying with longbows and a ballista, and have also dug a tunnel to hollow out the ground beneath the siege-tower. The excavation has been filled with vertical props. When these are set on fire and burn through, the whole siegework will come crashing down. Will the besiegers' counterminers be in time to kill the enemy sappers and prevent this collapse?

southern approach where secret paths up the cliff afforded some resupply. The resulting famine and disease that afflicted the Jews, now cut off from even meagre supplies of food and health-giving herbs on the rocky slopes, was intensified by the violent jealousy of the rival factions. Meanwhile, anyone who tried to escape – or even desert to the Romans – was crucified as soon as they were caught by the legionaries.

Titus evidently decided that Herod's fortress-palace on the west was too tough a proposition, so he ordered up more siege-towers and had them concentrate against the northern courts of the Temple and the Antonia Fortress, while his tunnellers undermined their walls. Simultaneously, the Jews were building a wall to act as a switchline within the Antonia. Then, quite by chance, the Roman tunnel collapsed; it had encountered an old counter-mine previously dug and abandoned by the Jews. The resulting cave-in brought down part of one of the Antonia towers, only for the legionaries to find themselves facing the new interior walls. Hurriedly improvising a ramp out of the tumbled masonry and rubble, the Romans stormed the new wall, were driven back, regrouped, counter-attacked, and conquered the Antonia Fortress. Without pausing, they tried to rush the Temple, but after a 12-hour battle they found themselves still confined to the Antonia.

Several times, Titus appealed to the Jews, either to surrender or to choose another battleground to preserve their Holy Temple. This the Jews regarded as a sign that the Romans were worried about attacking such a strongpoint. They rejected the appeal, assembled their remaining catapults on the Temple walls, and ignited the buildings outside the Temple proper, to deny cover to the attackers. For six days and nights the

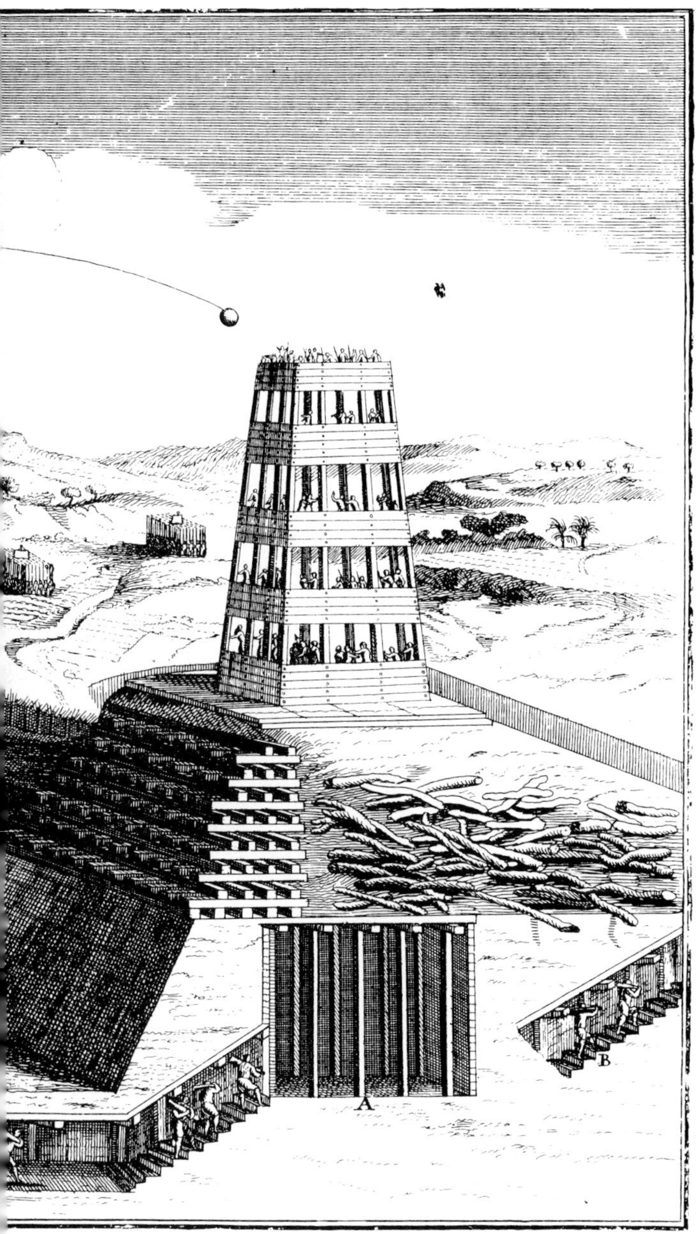

Romans battered at or tried to tunnel under, the Temple wall, but made little impression on either the facing stones or the inner core. Nor did scaling parties meet with greater success; those who reached the top were either hacked to death or thrown to the stone slabs far below. The Jews followed up their advantage by rushing out, destroying the nearest siege equipment, hurrying back and closing the gates again.

Next the Romans set fires against those gates which, though covered with silver plate, were of wood beneath. The metal melted, the timber charred, burst into flames, weakened, and disintegrated. While the blaze spread to adjoining woodwork and then into the outer buildings of the Temple complex, the Romans charged in, hunting down the Jews still holding out, until Titus ordered the legionaries to concentrate on extinguishing the fire before it reached the most sacred parts.

CADBURY CASTLE

CADBURY CASTLE is located about 10 km (7 miles) northeast of Yeovil in Somerset, England. Originally a Neolithic and Bronze Age settlement, it was fortified in about 500 BC during the Iron Age. The first stockade was an integral part of the rampart's wooden retaining wall, the latter subsequently replaced by stone facing. Between 250 and 58 BC, a further ditch and rampart were dug together with more complicated entrances, turning Cadbury into a multivallate hillfort to cope with the increased range of the new weapon, the sling. Within the ramparts were huts, granaries, workshops, and room for livestock. The hillfort was the political, commercial – and probably religious – centre for the surrounding rural community, who sheltered inside its fortifications in an emergency.

Cadbury Castle offered no resistance to the Roman invasion of AD 43-44. But some 30 years later Cadbury seems to have become the headquarters of either a freedom movement or a bandit gang. Whichever it was, the Romans wreaked fearful punishment, smashing down the gates and massacring the inhabitants amongst the ruins of their market stalls. Cadbury was abandoned for close on 500 years, until its defensive situation attracted the attention of a Dark Age warrior who may well have been King Arthur himself.

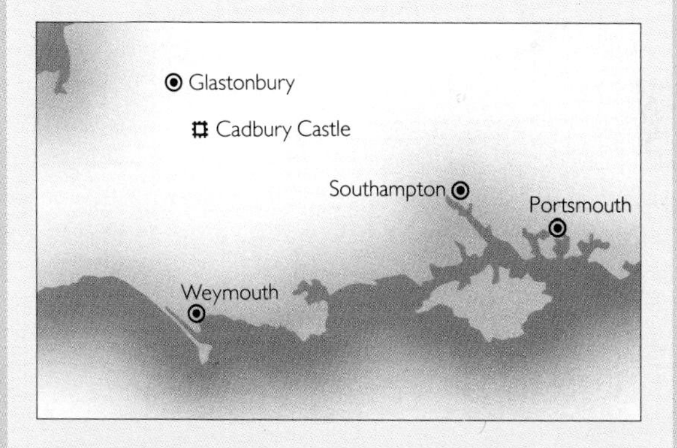

The south-eastern ramparts of Cadbury Castle, its location shown above.

From an attacker's point of view, the very walls of Jerusalem have eyes that watch his every move.

Titus was determined that the Temple should be preserved, even though that meant his men having to suffer heavy casualties as they penetrated the remotest, darkest recesses of the labyrinthine Temple in search of the last desperate Zealot. It was not a policy that appealed to the legionaries. One, thoroughly disenchanted with the length of time the conflict had raged, scrambled up on to a comrade's back and shoulders, and thrust a burning brand through an otherwise inaccessible opening, into an inner apartment – and the whole Temple was engulfed in fire, along with its last defenders.

Meanwhile the Roman catapults were concentrating on the First Wall until at last that was breached, and the legionaries swarmed through to seek, burn, and destroy. It is estimated that 11,000 people were killed in Hierosolyma between March-April and 8 September, AD 70. Another 97,000 were sold into slavery, although this must be the figure not only for the city but for the whole of Judaea.

Yet it was not the end of Jerusalem, nor indeed the total extinction of Jewish life in the city. Titus did, however, go on to order the dismantling of all its fortifications, except the towers of Hippicus, Phasael and Mariamne. A few Jews managed to remain in the ruins, and they were joined by more. Eventually sufficient numbers rebuilt their lives and their homes for Hierosolyma to start again to become a focus for Jewish dissidence. The result was another rebellion that lasted from AD 132-135, although Hierosolyma itself was stormed by the legions of Julius Severus in 133.

This time, Jews were absolutely forbidden to live in their Holy City, which was demolished and completely rebuilt, somewhat to the north, approximately where the former Lower and New Cities had been situated. Its wall followed the approximate line of the former Third Wall, with the partly-dismantled three towers of Herod's palace forming a massive strongpoint. Hierosolyma was renamed Aelia Capitolina, in honour of the reigning Emperor Publius Aelius Hadrianus.

The Romans did introduce something new into the overall strategic influences affecting military architecture: the concept of the entire state as one vast fortress. They did not think their way through to this doctrine nor did they necessarily themselves envisage the empire as one huge camp, but their frontier policy

SAXON SHORE FORTS

DURING THE third century AD, the main base of the *Classis Britannica* at Dubris (Dover) proved insufficient for the exercise of Roman sea-power in British waters by short-ranged galleys. Consequently, several forts were specially built on both sides of the channel and southern North Sea. Each one was surrounded by a stone wall. Projecting round bastions provided mountings for catapult artillery.

Each fort had to be big enough to accommodate sailors (including rowers), marines (for amphibious operations against pirate bases), cavalry (to round up any raiders who came ashore), artillerymen (for local defence), and all the technical and administrative staff, plus their workshops, stores and other buildings. There were also adjacent beaching hards and jetties. All the forts were under the command of the *Comes Litoris Saxonici per Britannicas* (The Count of the Saxon Shore). This title may refer to the area most likely to be raided by Saxons, or perhaps to an area that was garrisoned by Saxon mercenaries.

It was an early realization of the fleet-forts-field army strategy that has recurred several times in British history.

Burgh Castle, a Roman-Saxon shore fort.

THE ROMAN *LIMITES*

At the boundaries of the Roman Empire, sophisticated chains of forts were erected to keep the barbarians at bay

In AD 83-85, Domitian's legions invading Germany were ordered – as a temporary measure, pending the solution of a crisis farther east – to halt wherever they happened to be and erect a series of suitably-sited watchtowers, linked by a track through the forest. This was to run around a northerly salient from near Saalburg on the Taunus Ridge to Butzbach (near Giessen), and then south half-way across the Main valley – a total distance of 50 km (30 miles). Later, the advance would be continued.

On the flat land of Scythia Parva, however, Domitian's aim was simply to keep the barbarians out, not to invade their territory. So he ordered a long earth bank to be built stretching the 50 km (30 miles) from the Danube to the Black Sea. In some places the Dobregera *limes* still stands 4 m (13 ft) high. It has been realigned during its existence; sections pass both north and south of present-day Constança. (The latter line is sometimes called Trajan's Wall after the emperor who conquered neighbouring Dacia in AD 101-106.)

In Domitian's time some of the German watchtowers may originally have been linked by a light fence. Domitian abandoned his German ambitions and ordered that all the watchtowers be linked in this way. Meanwhile, the ends were extended to reach the Rivers Lahn and Main in the north and south respectively. This made a total length of 120 km (75 miles). The legions were ordered to remain on the defensive,

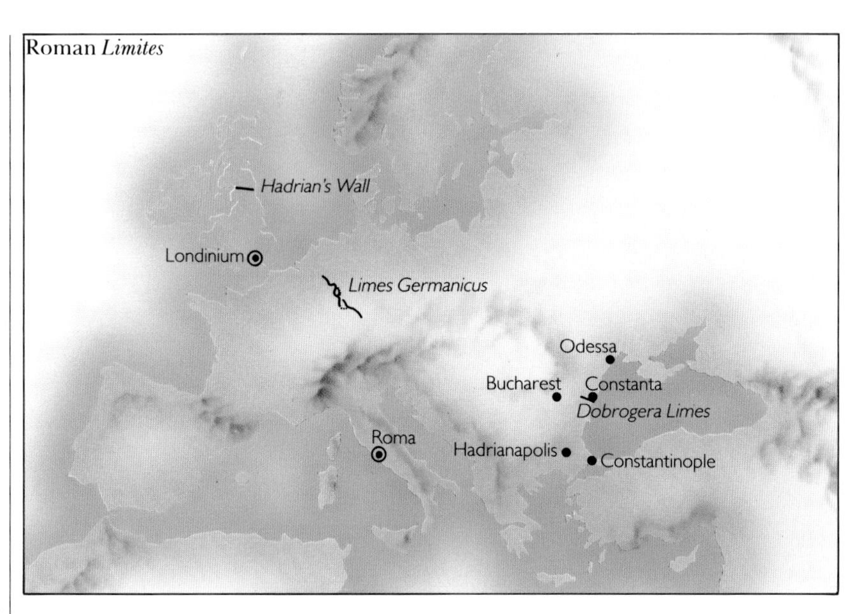

advancing further only on reconnaissance missions.

Hadrian (Emperor in AD 117) personally inspected the *limes Germanicus*. He ordered its realignment and reconstruction, and gave instructions for the building of an eastern section to begin at Miltenberg and to run south to the Neckar and then east to meet the Danube upstream from Regina (Regensburg) – a total length of about 280 km (175 miles). Some 55 auxiliary forts were to be established along the line of the *limes*, plus a further 30 forts in reserve. An additional road system was to be constructed to ensure legionary reinforcement from garrison towns in the rear.

This was the sort of linear defence-in-depth that Hadrian envisaged for the northern frontier of Britain, where there had been a number of attempts to conquer Caledonia. As a result of his tour of inspection there in AD 122, he ordered that a wall be built. Squared turves were originally used for one section, but in its final form Hadrian's Wall was of stone and ran from Wallsend to Bowness, a distance of 80 Roman miles (117.6 km or 73 English miles). It varied in thickness from 2.3-3 m (7½-9¾ ft). It was 4.5 m (14¾ ft) high with

Below: Just south of Hadrian's Wall itself – at Vindolanda – is this reconstruction of two lengths of the wall with towers: timber and turf on the left; stone on the right.

a crenellated parapet 1.5 m (5 ft) above that. It was garrisoned by *castra stativa* supplying troops for *castella* every Roman mile, with turrets in between. The *Vallum* (a ditch and rampart system) marked the southern limit of the military zone and inhibited access from that direction – Hadrian's Wall was as much to keep people in as to keep barbarians out.

It certainly prevented the Picts from learning anything of Roman preparations for a successful major offensive into Caledonia in AD 130-142. The *limes* was at that time established on the less substantial Antonine Wall, constructed to create a similar barrier along the 60 km (40 miles) between the Forth and the Clyde.

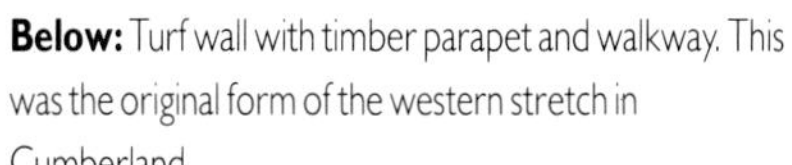

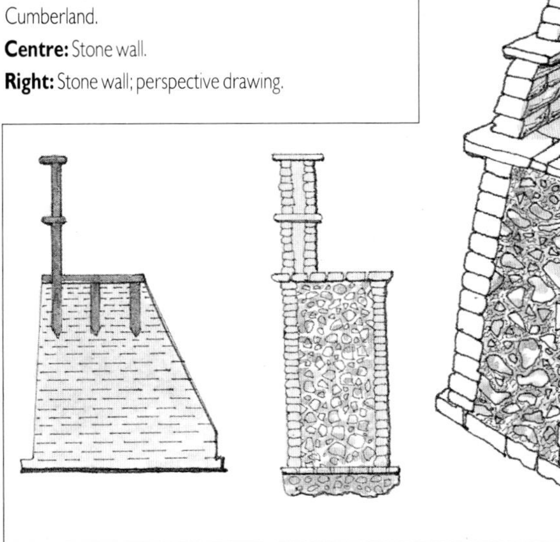

Below: Turf wall with timber parapet and walkway. This was the original form of the western stretch in Cumberland.
Centre: Stone wall.
Right: Stone wall; perspective drawing.

Meanwhile, over in Europe the *limes Germanicus* was also undergoing modification between AD 148 and 161. It seems that Hadrian's initial programme was never completed. Some of the wooden watchtowers on the Miltenberg-Neckar *limes* had indeed been rebuilt in stone. But Antoninus decided to abandon that line and, retaining Miltenberg as its northern hinge, ordered a new barrier to be constructed running more south-southeast to meet the eastbound line at Lorch. Stone watchtowers were erected every 500 paces, and the planned palisade of solid oak on top of a rampart was completed. In some places it was built or revetted with stone, the eastern section being known as 'The Devil's Wall'.

Back in Britain the Antonine Wall was abandoned during a rebellion by the Brigantes in 158-165. Hadrian's Wall was recommissioned first as the northern frontier, and then as the linear base for the reconquest of Caledonia and the reoccupation of the Antonine Wall. In 197 Clodius Albinus shipped the entire British garrison to Gaul in his bid to become emperor. Both the Antonine and Hadrian's Wall were overrun, but when Severus became emperor, the barbarians withdrew and Hadrian's Wall was rebuilt as the northern *limes*. In 296 the Roman army was again taken overseas as a pawn in political manoeuvrings – and again the barbarians stormed in. Once more, however, they retreated before the return of imperial government.

In 367 Hadrian's Wall was betrayed to the enemy in what is sometimes called the Barbarian Conspiracy. Again rebuilt, its defence was entrusted to local militia. Many of the troops would have been ex-legionaries and auxiliaries, quite capable of instructing new recruits and of organizing a successful defence. But with the break-up of the Roman Empire – as much due to local independence movements as to foreign invasion – Hadrian's Wall ceased to have any significance, except as a simple boundary and as an easily workable source of building-stone.

Hadrian's Wall looking east from Hotbank Crags near Haltwhistle. A romantic scene, the wall was once the limit of civilization, a military zone forbidden to unauthorized personnel, a springboard for northward invasion.

seems nevertheless to have been an instinctive extension of their tactical experience and practice.

THE *CASTRA STATIVA*

Every single day that a Roman army was on campaign, the march was halted early enough to enable trenches and ramparts to be dug and a palisade erected, behind which the unit was secure during the hours of darkness. In the morning the tents were struck, the stakes pulled up, and the rectangle of ditches and banks abandoned. This temporary encampment was called a *castra*, but if several nights were to be spent in the same place, then it was designated *castra stativa*. It was strengthened with stone walls, wooden stockades and *castella* (elevated and projecting redoubts).

Castra stativa served as bases for operations in the surrounding countryside. The legionaries left their baggage (*impedimenta*) in the camp and went out *expeditus*, each man wearing and carrying only his combat gear. If the engagement did not go favourably, the legion withdrew in good order, recoiling upon its reserves of food, ammunition and reinforcements in the camp. And if necessary, it was comparatively easy to turn the *castra stativa* into a permanent fort by re-

building everything in stone.

Forts of this standard layout were established in the frontier zones of North Africa, Arabia, Asia and the Balkans. They fulfilled the same function as the ancient Egyptian *migdol*, simultaneously bases for wide-ranging cavalry operations and obstacles to barbarian advance along the few highways through wilderness or mountain.

The Saxon Shore naval bases established by the Romans on the coasts of Britain and north-west Gaul from AD 220-230 onwards were the maritime equivalent of *castra stativa*. However, in some places there were no natural barriers either to keep the barbarians out or to funnel them into battlegrounds disadvantageous for them. So, on the Dobrogera Plain from the Lower Danube to the Black Sea, over the mountains and valleys between the Upper Rhine and the Upper Danube, and through the barren land in northern Britain, walls were built.

From behind these walls the legions marched out and the cavalry *alae* ('wings') rode out first to confront the barbarians and annihilate them before they reached the frontier lines (or *limites*), and second to lay waste and depopulate their homeland so that the border was quiet for another generation. Occasionally, as a result of temporary government incompetence or intrigue, the imperial defences fell below strength. Then the local tribes made common cause with their kinsfolk beyond the wall, seized the gates, opened them and let in the barbarians. For a while the law-abiding citizens of the province knew terror – until fresh legions arrived and the empire was safe again.

There finally came a time when the metaphoric walls of the empire really did collapse, as much through local independence movements as by foreign incursion. And then the real walls were abandoned. They had served their purpose for four centuries. It would be almost twice as long before another form of permanent work established itself as the dominant type of European military architecture.

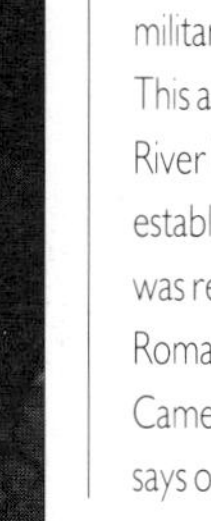

Civilian amphitheatres may have staged wild-beast shows, gladiator contests and Christian persecutions. Army ones were more usually employed as parade-grounds, exercise areas and arenas for military displays (what we call 'tattoos'). This amphitheatre is at Caerleon on the River Usk in South Wales. First established in the mid-70s AD, Caerleon was rebuilt about AD 100. After the Romans left, it became the site of Camelot, the court of King Arthur – or so says one legend.

Chapter Two

THE MEDIEVAL WORLD

The most serious loss resulting from the end of the Roman Empire was the collapse of the idea of the world state. Europe disintegrated into a multitude of petty kingdoms, jealous cities and suspicious villages. It was a time of genuine self-sufficiency. If any community could not make or grow what it needed, they went without and died – or stole from their neighbours.

Big communities – like cities – refurbished their ancient walls, generally under the instructions of some retired warrior.

Small communities – like remote hamlets – were simply abandoned when danger threatened: the inhabitants hid in the woods until the danger passed on. Alternatively, frightened peasants might ask the owner of the nearest farmstead if they could shelter behind his stout stockade. They might find grudging acceptance there – if he had enough food put by.

Gradually, over the decades, this second arrangement became the norm in many areas. The peasants lost what independence they may have had and became serfs and labourers, working on the farmstead-owner's land and subject to immediate conscription if the property required defending. At the same time, the farmstead-owners attained the status of chief or lord, able to take decisions that affected many lives – such as whether to repair and refurbish the old Roman watch-tower on the hill near by, or whether instead to rebuild

Josselin (**left**), rising conical-towered like a Walt Disney fairytale castle and lily-moated Bodiam (**right**), are exactly how medieval castles are imagined. Both, in fact, are quite untypical. In 1370, the original castle of Josselin on the River Oust, was made into a French royal bastion against Breton nationalism. The Duke of Brittany dismantled Josselin in 1488, but later it was turned into the palace we see today. Its remaining fortification was blown up on Richelieu's orders in 1629. Conversely, multi-styled Bodiam in English Sussex, was actually built in 1385 to take advantage of the new gunpowder weapons.

Left: According to German legend, a cave in this cliff (Drachenfels), overlooking the Rhine, was the lair of a dragon guarding the treasure of King Nibelung. Siegfried killed the dragon, bathed in its blood, and stole the treasure. In historic times (in 1117), the site, 1050 feet (315 metres) above the river, was fortified. The present castle (now ruined) was built by Archbishop Arnold of Cologne in 1147.

Right: Just a few of the 88 round towers providing fire along the front of the curtain walls of Avila. In 1090, King Alfonso VII of Castile ordered his son-in-law (Raymond of Burgundy) to develop the city as a base of operations and a centre of culture in central Spain. By 1099, the architects Cassandro and Florian de Pontieu and 2,000 expert stonemasons (plus an army of labourers) had built a fortress with a defensible perimeter of 1¾ miles (4 kilometres). The work was speeded by incorporating existing Roman walls. The towers resemble the rounded bastions of Saxon shore forts. Some of the merlons are pointed, an Arab feature reflecting the employment of Muslim masons (as well as Jewish and Christian). Avila was the first medieval European city to have drum towers, and the first to be walled on such a scale.

the old prehistoric hillfort on the other side of the valley. There, with proper gates and walkways, an enemy could be kept out for ever; and there would be ample space for herds and granaries and stables and a feasting-hall for his warriors, a workshop for his armourers, a chapel for his priest to pray in, and even somewhere for the peasantry to huddle.

The peasants found that they had exchanged one anxiety for a host of others. Could they meet their lord's demands for labour and corn – or would he take *all* their food as callously as any foreign thief? Would they be able to reach the refuge before being overtaken by marauders? It was no good now trying to run off and hide in the woods, for now the very fact that they belonged to the lord of the area made them legitimate targets for the tax-gatherers and swordsmen and lawyer-priests of all the other lords who also claimed sovereignty over the same area.

THE END OF THE DARK AGES

All the same, it is unfair to portray the so-called Dark Ages as an era of unremitting rapine and murder. In some places a whole generation could live out its life without disturbance. And in other areas, minor skirmishes were magnified into pitched battles and bloody massacres by balladeers and priestly propagandists seeking to bolster the ego of their own particular overlord. Trade was carried on and knowledge transmitted to a certain extent.

Above all, there was the doctrine of the ideal unity of Christendom, reinforced after Christmas Day AD 800 by the establishment of the rank of Holy Roman Emperor. It was an attempt to revive the law and order of a lost imperium 400 years after it had collapsed.

What might have been a great unification movement, however, had by 1100 provided further opportunities for discord, as rival kings and regents, dukes and warrior-prelates, sided with either Pope or emperor, always according to their own self-interest, but always claiming that right was on their side.

Impelled by self-preservation and religious certainty, many of these protagonists built themselves stone strongholds on inaccessible crags, from where they could survey their principality far below, and from where they could sally forth to assault their neighbours' territories. Other nobles withdrew into fortified centres of trade, industry and administration, the planned construction of which had been part of Henry the Fowler's imperial policy for defending eastern Germany against the Magyars and Slavs in 918-936. These cities were known as *burgen*. The Germanic Saxons who crossed the sea to Britain used their

NORMAN CASTLES

The stout walls of the Norman motte and bailey were well designed to keep subject peoples in check

Berkhamsted Castle, in Hertfordshire, is a near-perfect example of the motte-and-bailey castles erected by the Normans when they came to England. The bailey which measured 137 × 91 m (450 × 300 ft) is completely surrounded by a bridged ditch which floods in winter; an outer ditch is a later addition. Separate – and only accessible via the bailey and another bridge – is the conical motte 55 m (180 ft) in diameter, tapering to 18.3 m (60 ft) at a height of 13.7 m (45 ft). The foundations of the round stone keep are believed to reach right down to the chalk, which suggests that it was built as a tall tower and the earth motte piled around it.

So Berkhamsted was probably a stone castle from the very beginning. The first castellan was Duke William's brother, Robert, Count of Mortain, who owned this land from the earliest days of the Conquest. Another of his conical mottes can be seen completely isolated at *Mons Acutus* (Montacute) near Yeovil, in Somerset. Now topped by a stone folly, in Norman times this artificially-shaped natural hill bore a small wooden fort. Yet another of the Mortain natural motte strongholds was Launceston in Cornwall. There was once a bailey, although it is not now readily apparent – only the ruins of two concentric round towers remain; the outer lower one was the original shell keep.

Not so obvious as a Norman motte-and-bailey castle is the castle at Windsor, Berkshire, now famous as a royal palace. But within the

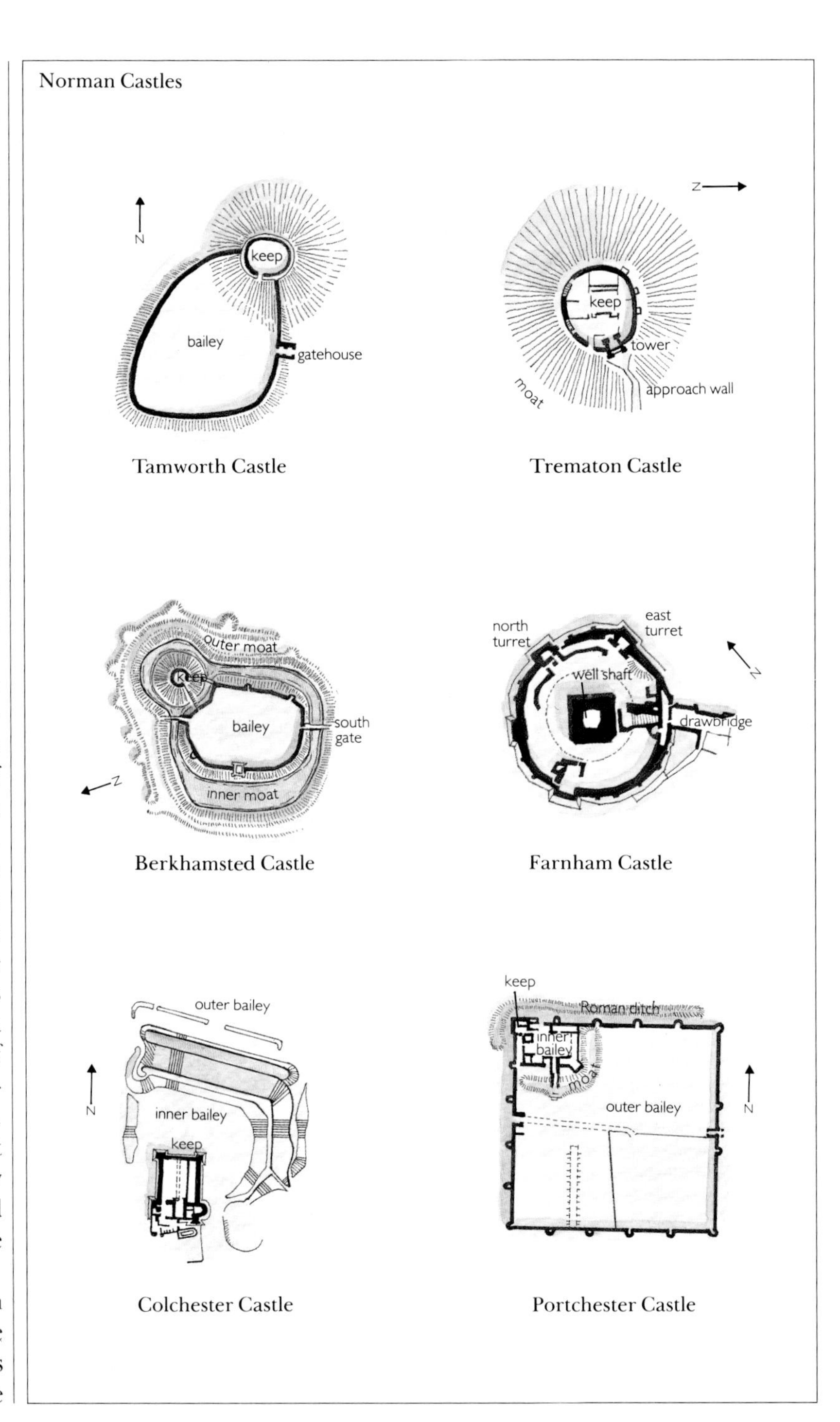

later curtain walls of the Upper and Lower Wards, the Round Tower rises from a steep-sided but natural chalk hillock. The lowest stonework is the remains of Henry I's shell keep, which replaced William I's original timber fort.

Above: The motte raised to support William I's timber castle at York in 1068-9. Later this stone keep of quatrefoil plan was built. It derives its name of Clifford's Tower from the rebel baron whose body was hung in chains here in 1322.

Rochester Castle (**above**) is an example of a square donjon, while the Bayeux Tapestry (**below**) shows how such buildings were depicted at the time.

A variation on the motte-and-bailey theme, but even less apparent, is illustrated by Portchester Castle in Hampshire. Here the Outer Bailey is formed by the walls around a Roman Saxon Shore fort. But instead of raising a mound of earth in the northwest corner in the 1120s, Henry I constructed an Inner Bailey: a bridged wet ditch backed by a stone gatehouse and curtain wall, within which is a donjon. A stepped plinth forms the foundation for the walls and six pilaster buttresses which recede into the four corners and two exposed faces of the square keep.

The most famous square donjon in England is the Tower of London. It seems likely that William I built a wooden fortification here. Like Portchester, its Outer Ward was bounded by a Roman wall. The foundations of London's later great stone tower were laid for the king by Gundulf (soon Bishop of Rochester) in about 1076, being completed by about 1097. The walls of Her Majesty's Tower rise from a sloping plinth (or batter) over 30 m (100 ft) square. They are 4.6-3.4 m (15-11 ft) thick and reach a height of 27.5 m (90 ft). A massive internal wall from top to bottom partitions the Tower into two defensible fortresses. Three of the corner towers are square; that on the northeast is round.

Henry III (ruled 1216-1272) began the outworks which eventually transformed the castle into a concentric fortification. He also ordered the donjon to be whitewashed inside and out for cleanliness and preservation; it has been the 'White Tower' ever since.

In 1087-1089 Gundulf began construction of a square stone castle for the King at Rochester, Kent. The donjon was completed by Corbeil, Archbishop of Canterbury. Like the Tower of London, one of its corner towers is round – but it was not initially constructed like that. It was rebuilt in a round style after being destroyed during the revolt by the barons against King John in 1215.

It is interesting that of these seven fortifications, three round keeps belonged to Robert of Mortain and three square donjons to the king – round Windsor being the final exception. Yet it may be that in spite of all the debate about the merits and disadvantages of square or round towers, all that decided their shape was the nature of the site on which they were built or simply personal preference for a certain type of fortification.

cognate word *burh* to describe those particular urban strongholds first noted in the reign of Alfred the Great (ruled 871-899).

No English town could be the location of a royal mint without being surrounded by a palisaded earth bank or a parapet-topped stone wall of military significance. These places also served as headquarters for cavalry operations against Scandinavian horsemen sweeping across country from their own landing-places.

Although most renowned for their maritime wanderings, the Vikings themselves appreciated the importance of secure bases for warships and warriors. If, late in the season, they found themselves far from home, they tried to capture a walled town for their winter-quarters. If that failed, they hauled their boats ashore, turned the unserviceable ones upside-down as improvised huts, and encircled the camp with rampart and fence. When spring came and the squadron departed, some of them might remain to develop a temporary camp into a permanent settlement.

THE NORMANS

Vikings – Danes, Swedes, Norsemen (or Northmen) – they ranged the known world. They sailed the western Ocean perhaps as far as America. They conquered Sicily. Under Rurik they colonized some of the harsh lands to the east of the Baltic. Some settled in northern Gaul and – as Normans – waged a long-term struggle for existence and aggrandisement against their neighbours, among whom was Fulk Nerra.

Fulk the Black was one of several claimants for the fertile valley of the Loire. It eventually fell under his dominion thanks to a series of stockaded forts. Each one served as a cavalry base for the ruthless exploitation and administration of the surrounding area, and as a springboard for the conquest of the next patch of farmland. Fulk had inflicted fire upon other people often enough for him to appreciate the destructive power of that element, and to be aware of the defensive advantages of a stone fortress. Accordingly, in about 995 he ordered the construction of one at Langeais, west of the Count of Blois' city of Tours. It probably followed the pattern of Fulk's earlier timber strongholds.

Langeais was a simple rectangular tower about 16 × 7 m ($52\frac{1}{2}$ × 23 ft) in plan. Its buttressed walls were of smallish, roughly-hewn masonry. Entrance was by means of a small turret, which projected in such a way that anyone approaching the door would have to expose their right, unshielded side to missiles from the top of the walls. Once inside, an attacker had to

The old walls of Mksheta, near Tbilisi in Soviet Georgia. This stronghold is situated in one of the narrower parts of the valley linking the Caspian and Black Seas, and separating the Caucasus range from the mountains of Asia Minor. In the late Middle Ages, the independent state of Georgia (known to its people as Karthveli) was thus a bastion of Christian defence against Tartars from the east, Turks from the west, and Arabs and Persians from the south. The last king (George XIII) resigned his kingdom in favour of Tsar Paul and in 1799 Georgia became the Russian Gruzia.

climb up a stone staircase to reach the first floor above ground level, where the entrance was located to the east living quarters of the castle. The ground level and basement were storerooms, without windows for greater security.

Because the Dukes of Normandy shared a common boundary with the Counts of Anjou, it was inevitable that they too should refurbish stone defences in their territory, and construct new fortresses of stone both for practical military purposes and to symbolize their own proud strength. And it is with the erection of stone castles by the Normans that the great period of Medieval strongholds begins.

The first military architecture that the local English people saw when the Normans arrived was very definitely temporary in nature. The knights and their retainers were accompanied by waggons and pack-horses carrying not only food but also stout wooden panels and posts. Each night when on summer campaigning, their servants enclosed a small area with ditch and bank, dug postholes, and hammered the prefabricated panels into the ground to form a continuous gated stockade. It was not an impregnable fortress, but once the enemy's field army had been defeated (as at Hastings), it was certainly adequate against a disgruntled peasantry. Next morning it was all dismantled, restowed, and the column set out again.

More substantial measures were taken when a Norman baron first arrived in the territory in which he was going to live. The local peasantry were rounded up and set to work steepening and smoothing the slopes of a natural hillock standing on the spur of a low ridge, which could itself be similarly improved. If no such suitable geological formation presented itself, then the labourers had to dig a circular or rectangular ditch. The excavated material was piled up in solidly compacted layers until a conical mound had been formed. This artificial hill was called a 'motte'.

Sometimes surrounding the motte, but more usually in front of it, was a large cleared area, itself enclosed by ditch and rampart. This space was called the 'bailey', where the lord's chief servant (originally his armour- and sword-bearer, in Latin *bajulus*) was in charge.

Meanwhile other teams were despatched to cut down trees, haul them out of the forest and trim them, for the erection of a palisade around the bailey and for the construction of a wooden fort on top of the motte.

This building's title of 'donjon' is derived from the same root as the Latin words *domus* for 'house' and *dominus* for 'lord'. In due course this term was transferred to the cellars *beneath* the house, the 'dungeons', while the word motte was transposed to mean the ditch or 'moat' *around* the mound. The donjon could also be referred to simply as 'the tower', or at other times as a 'keep'.

CONSTANTINOPLE

In AD 324 Constantine chose Byzantium as the site of his 'New Rome'. Within 6 years it had been completely rebuilt, enlarged, and refortified; further expansion necessitated new fortification under the Emperor Theodosius II.

The land walls ran 5 km (3 miles) from the innermost creek of the Golden Horn inlet on the north shore to the Sea of Marmora in the south. The main wall was 4.6 m (15½ ft) thick and 27 m (88 ft) high; 55 m (60 yd) in front was another wall, 2 m thick and 5.5 m high (6½ × 18 ft). Beyond another terrace fronted by a battlemented parapet, came the outermost defence: a ditch 18 m wide and 6.5 m deep (55 × 19 ft), floodable and containing a variety of obstacles. The two main walls were each provided with 96 projecting towers (on the inner) and bastions (on the outer). Both were capable of mounting the heaviest catapult artillery and were so located that the outers did not mask the inners' field of fire. Each gateway was a mighty fortress in itself.

For more than a millennium these defences held against all assailants.

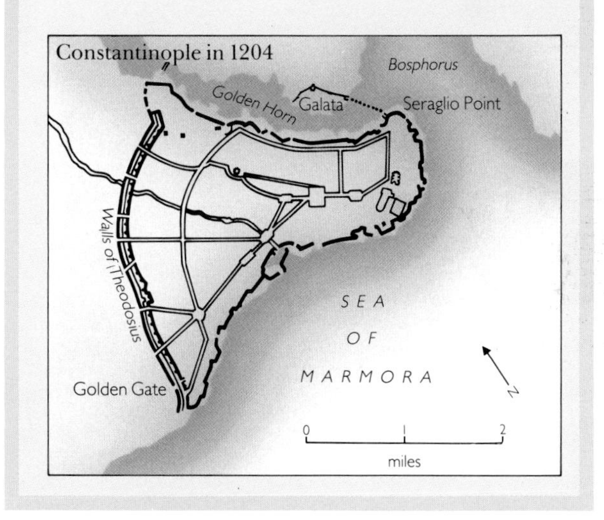

The keep was reserved for the Master, his family and his immediate entourage, while the lord's other servants, his soldiers and skilled artisans lived in huts in the bailey. This also served as an exercise area, a parade ground, a corral for stock in an emergency, and – sometimes – as a market square. Lesser servants dwelt in their own hutments right outside the complex, coming in daily or whenever their labour was required. Sometimes they were allowed to shelter in the bailey during a siege, but that was a boon, its dispensation depending upon military necessity and the social conscience of the castlellan.

Once the wooden motte-and-bailey castle had been completed, the baron had other things to think about.

Above: Cairo Citadel, begun by Saladin in 1176.

Left: Beaumaris is a perfectly concentric castle on the island of Anglesey, archers on the inner bailey firing over the lower outer walls. It was built as part of Edward I's pacification of North Wales, although not completed until 1320.

There was the question of his security of tenure, mainly consequent upon his relationship with his own overlord. How fast was his timber fortress rotting? How much coin did he have? This last was a most important consideration in any decision to improve the fortifications. Unless his territory included a suitable quarry, the baron would have to import both stone and the skilled masons to work it – plus the scaffolders without whose unlettered expertise no substantial building above ground level could be constructed. Such craftsmen and merchants were independent; they were not owned by the baron; they could not be forced to labour as part of their bounden duty; nor would they be obsequiously grateful for a meagre dole of bread, cheese and ale. No,

they demanded gold and silver coins, and unless the baron had some, he would get no stone castle. But a baron's prestige as well as his very existence depended upon his having a stone castle, and eventually the work would be begun.

Usually a fresh site was selected, partly to enable the old wooden fort to continue in use undisturbed during construction, and partly because the massive weight of a stone castle required more substantial foundations than could be provided by an artificial pile of earth, no matter how compactly it had been laid down. Accordingly, most stone castles were built either on rocky outcrops or where solid substrata were very close to the surface. In both instances, the result was the same: the castle was located on a natural eminence. Thus local topography assisted defence, and the castellan's self-esteem was enhanced by his elevation above his ground-level subjects.

Such a lofty viewpoint also proved to have other practical considerations. Castles not only overlooked nearby roads and bridges, their missiles also prevented all movement along those routes. A hand-drawn longbow had a range of 200 m (220 yd); the cranked-spring crossbow could project its heavier armour-piercing bolt some 350-400 m (380-440 yd) in the hands of a skilled marksman. Catapult artillery had a similar range of about 400 m (¼ mile), both distance and striking-power enhanced by gravity – the same natural force, a disadvantage for archers down at ground-level, firing upwards.

Nor could an invading army try to hurry past, accepting casualties but otherwise ignoring the fortress. There might, after all, be a vast number of knights and men-at-arms in the castle, ready to fall upon the rear-guard or – worse still – to bar the line of retreat if God did not favour the invaders' enterprise. So the castle had to be captured, preferably by immediate storm. The defenders were sitting on a great store of food and munitions. The invaders had to bring everything with them, or else waste time and effort foraging the countryside and arousing the hatred of the locals who were already expert at concealing what little sustenance remained to them after their own lord's depredations. Inside the castle were warm hearths, dry clothes, hot meals, a comfortable bed at night, proper sanitation. Outside was rain and snow, mud and wet feet, stones beneath the sodden blanket, a quagmire of a cesspit, and the ever-present danger of attack by the besieged's field army. A long-drawn-out siege was out of the question. The repulse or success of an immediate assault would show whose side God was on in this dispute.

There were some exceptions to this chivalric conduct of castle warfare. No king could permit a rebellious baron to remain in possession of a royal castle. If he could not capture it at once, the monarch settled down for a long siege with his army – or sent mercenaries to keep up the siege while he himself attended to

Left: Berkhamsted Castle (north-west of London): the remains of the bailey wall, and the stepped causeway leading across the wet ditch and up the motte where the shell keep once stood.
Right: Octagonal Odiham Castle, near Basingstoke in Hampshire. Here, for fifteen days in 1216, three knights and ten sergeants withstood a French army of 7,000.

CADBURY CASTLE

In about AD 500 some unknown British leader established his stronghold in Cadbury Castle, a long-deserted prehistoric hillfort in Somerset. The topmost rampart was surmounted by dry-stone walling, with timber-reinforced earth-filling, and crowned by a wooden battlement; there was an impressive wooden double-gateway. The buildings within included a feasting-hall measuring 19.2 × 10.4 m (63 ft × 34 feet). The Dark Age splendour of this complex adds weight to the tradition that Cadbury Castle is the true site of King Arthur's Camelot – it is only one hour's hard ride from Avalon at Glastonbury.

In due course this Dark Age settlement was abandoned, but in the winter of 1009-10 King Ethelred ordered a royal mint to be established at Cadbury (or Cadanbyrig as it was then called), and a new wall on the topmost rampart was set up. It had a thickness of 6 m (20 ft), comprising a mortar-bound core of rock-rubble and a facing of lias slabs. The gateway was a long passage about 9 m (29½ ft) by 3 m (10 ft).

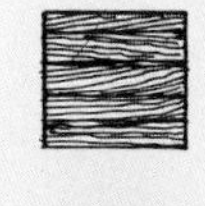
timber

stone

Early prehistoric

Later prehistoric

Early Dark Age (Arthurian period)

AD 1009/10–1019

other tasks of government – most of which involved castles anyway.

For medieval fortresses were not only military strongholds, they were also centres of feudal administration. In what was virtually the privatization of law and order, the king gave to barons castles and supportive manorial lands to enable them to enforce the law within their own territory, to collect taxes and, especially in border areas, to serve as the first line of defence against invasion. Here and there the king sited his own strongholds to act as bases for operations against overmighty subjects or against any foreign army that proved too strong for local levies and retainers. But for the most part the king was content to let his nobles do his work for him – provided they forwarded their due proportion of the taxes collected and performed

Edinburgh Castle: stronghold, seat of governance, palace, prison, museum, arena for military and cultural displays, site of St Margaret's Chapel, and home of the Scottish National War Memorial.

any other duties that were required as proof of their allegiance to him.

It was sometimes difficult, however, to be sure that allegiance was being pledged to the right person. In an unlettered society of complex marriage alliances and annulments, excommunications, minorities and regencies, even loyal barons had to remember that today's pretender to the throne might tomorrow be the true liege lord. The common people might not have realized it, but early medieval siege warfare was consequently conducted in a relatively gentlemanly manner, always allowing an opponent a chance to depart or surrender with dignity.

THE CRUSADER CASTLES

The Crusades changed all that, and the new attitudes were reflected in new developments in castle architecture.

The city of Jerusalem is sacred to Jews as the site of the Holy Temple; to Christians as the place of the Crucifixion and the Resurrection; and to Muslims as the spot miraculously visited by Muhammad prior to the Revelation of the Word of God. The city has changed hands many times during its history. Critically, in 637 it was captured by the Arabs during the initial expansion of Islam. Then in 1077 the Seljuk Turks (also Muslim) conquered it. 18 years later, Pope Urban II proclaimed a Crusade to win back the Most Holy Places of Christendom. What was preached was salvation. What was promised was adventure. And what practicality demanded was profit.

The result was a century of Mediterranean conflict – eight Crusades of varying efficiency – during which Jerusalem was captured, governed and lost, and Constantinople was captured, exploited and lost.

Whatever their personal aims on setting out, the Crusaders who reached and fought in the Holy Land soon realized that their overriding motivation had to be self-preservation. This was especially true of those Orders of Knights – warrior-monks – who were left to garrison the Kingdoms of Outremer ('Over-the-Sea') after everybody else had gone home. At first they lived a life of comparative peace and increasing comfort, either in captured fortresses of ancient foundation, or in newly-built or restored towers and citadels, usually of square Norman appearance similar to those in France or Sicily.

But from 1144 onwards the Crusaders were vastly outnumbered by Saracen armies whose leaders rekindled the missionary zeal of Islam and organized Turkish energies for war. For them, Crusader castles had become symbols of Frankish domination and infidel pollution. They could not be ignored or bypassed en route to a favourable battlefield. Nor could they be besieged, the defenders resisting stoutly and then surrendering honourably.

Each and every Crusader castle had therefore to be stormed and either destroyed or purified in some way, preferably with the blood of its inhabitants – unless they chose to become Muslims. And that was impossible for Crusaders dedicated to exterminating their foes. Both sides were equally assured of eternal life if they fell in battle.

So the castles of the Holy Land became sophisti-

JERUSALEM

Muslim Arabs captured the Byzantine city of Jerusalem in AD 637. Thereafter, the city (renamed Al-Kuds, or 'The Holy One') changed Muslim rulers several times, and it was not until 7 June 1099 that the First Crusade arrived outside its walls.

The first assault failed until the two chief engineers (Gaston of Bearn and William of Ricou) had constructed two huge siege-towers. They were fireproofed with leather hides sewn together by women and elderly pilgrims. In the thick of action, however, the Crusader towers started to shake apart, and the people on the top felt seasick with the continual swaying.

The Crusaders' attention was then directed to the north wall, where huge mattresses had been suspended from massive beams to cushion the masonry. With a siege-tower, one mattress was cut down and others were ignited by the Crusaders, the thick smoke blinding and choking the Saracens and driving them from the battlements. Dropping their tower's drawbridge, the Crusaders stormed the city, while the southern attack force scaled the undefended walls in their sector.

But on 20 September 1187 it was the Crusaders – in repaired fortifications – who were under siege with not enough men and no field army to bring relief. They tried to break out, but were driven back by the Saracens, who in turn broke into the city on 3 October 1187.

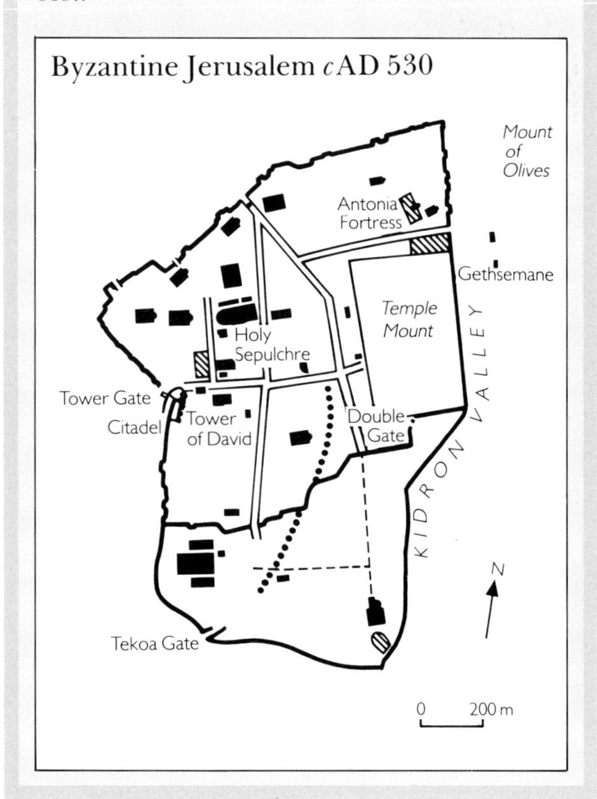

Pillars 6 feet (almost 2 metres) in diameter, support fan-vaulting in the Crusaders' Hall of the castle at Acre on the Palestinian coast.

cated killing-grounds, weapons systems designed to enable a few defenders to hold out for as long as possible while selling their lives as dearly as possible. Usually located on some inaccessible mountain-top, each massive keep was now encircled by two or three concentric walls, so arranged that the inner higher defenders could fire over the heads of their comrades on the lower outer ramparts. Each of the curtain walls

CRUSADER CASTLES

In the struggle against the Moors the Crusader castles were outposts of Christendom in a pagan land

The warrior-monks left to garrison the Holy Land after the successful First Crusade (1096-99) built the forts and fortifications that seemed best to them at that time and for that location, trusting in God and their own expertise to select the strongest site and make the walls impregnable. The result was an apparently haphazard collection of strongholds whose garrisons policed, taxed and administered the hostile peoples of the land in between. They were usually on mountain-tops for inaccessibility, to force besiegers to labour in an inhospitable environment, and for signalling through the clear air to the next distant fortress. The Franks tried to ensure that no castle could be directly attacked without its assailants having to pass through the territory patrolled by another, or perhaps along a road liable to withering fire from its own catapult artillery.

Of first priority – though invisible to outsiders – was the accumulation of reserves of food and water. Sahyun (or Saone) had stone cisterns that could contain 10 million l (more than 2 million gall), vital not only for men but also for the horses upon which a counter-attack depended. (Any garrison that was reduced to eating horse-flesh knew they had no hope of ever breaking out.)

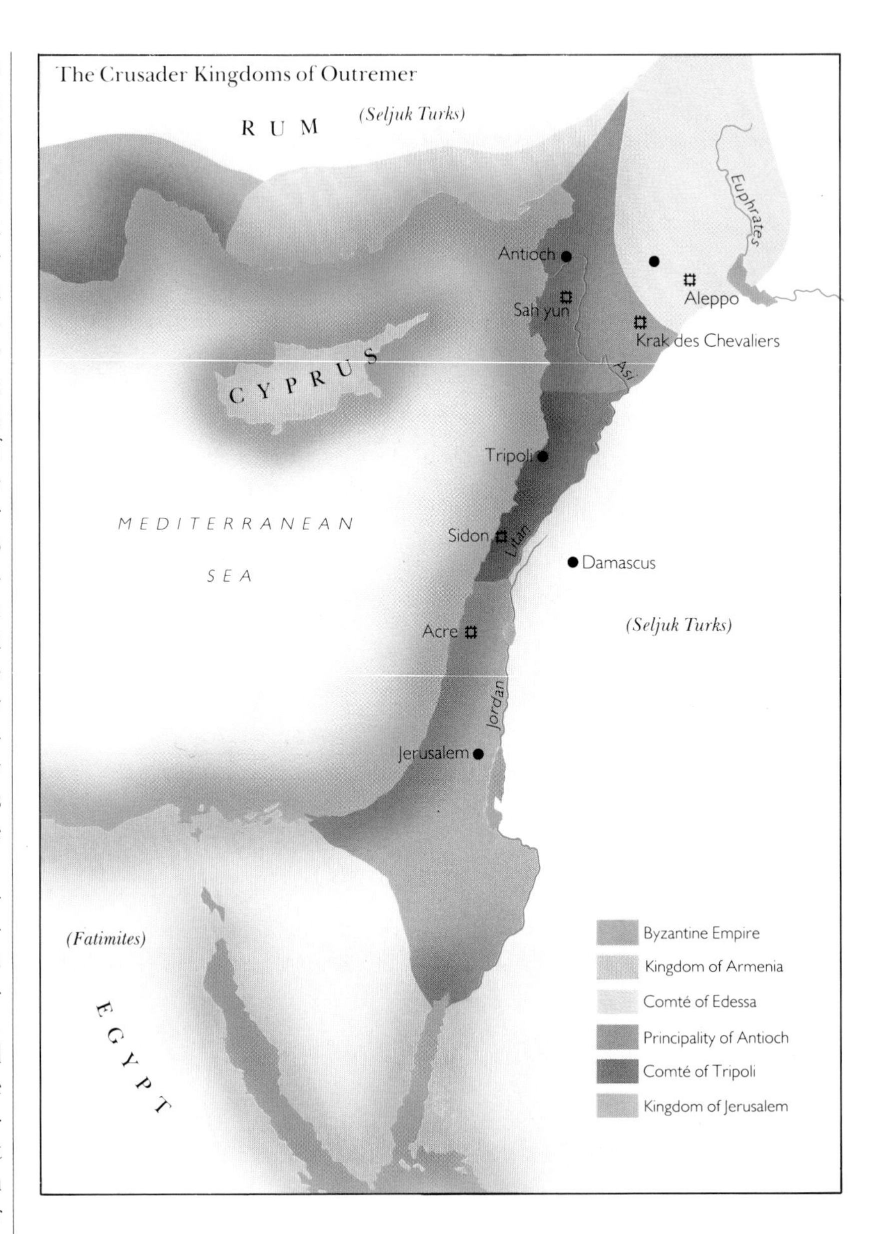

Sahyun guarded the southern approach to Syrian Antioch (in modern Turkey) and was built on the site of an ancient square Byzantine fortification, although it is much larger. Sahyun extends the full length of a 600-m (660-yd) spur. The Upper and Lower Wards are both surrounded by their own curtain walls and separated by a ditch. The only way into the castle is across an artificial canyon, hand-cut through the solid rock across the full 130-m (140-yd) breadth of the ridge. It is 27 m (90 ft) deep and too wide (19 m, 63 ft) to be spanned by a single bridge strong enough for horses. So the labourers left a slen-

The citadel at Aleppo (Haleb in modern Syria) was rebuilt by the Saracens after the Crusaders' departure. This photograph shows that even when robbed of its stone facing and eroded by rain, the glacis still provides no cover for attackers.

der pinnacle in the centre of the canyon to support the middle of solid yet movable decking. Dating from the early 1100s, most of Sahyun's towers are square in appearance.

Krak des Chevaliers (north of Homs in Syria) was also first fortified in the early 1100s, but was handed to the Knights Hospitaller in 1142; it is their alterations that account for most of the stronghold's present appearance. It is a concentric castle with an approach through four changes of direction and eight gates in three gateways – provided the correct choice is made at each complication. Three square towers (a fourth is later Saracen work) and nine round towers guard the outer wall perched on the cliff edge. The inner wall rises from an immense talus (or plinth). Here there are seven towers of varying types, including sally-ports for commando raids upon besiegers. Part of the Lower Ward itself formed a huge reservoir; there were other cisterns to collect the winter rains. Krak des Chevaliers also boasted a windmill. Grain could be stored longer than flour, but there must also have been container-storage of spices and sauces, sweetmeats and oils, dried fruits and herbs to relieve the monotony of the staple unflavoured, gritty pasta.

2,000 Knights Hospitaller put up such a defence for a month in 1188 that Saladin himself was compelled to abandon the siege and turn to easier targets. But in 1271 Krak des Chevaliers' troops, reduced by then to 200 warrior-monks plus serving men were too few to man the whole of the outer perimeter. Once in possession of the Outer Ward, the Saracens undermined one of the inner wall towers: the Hospitallers were confined to the keep, itself an integral part of the inner curtain wall. But with each retreat they found the reduced perimeter easier to defend. Sultan Beibar's force had suffered enough casualties in the previous three weeks for him to appreciate that storming the final redoubt would be far too costly. So he arranged for a Muslim of Frankish appearance and manner to pretend to be a messenger who had somehow got through the Saracen lines with instructions from Hospitalles authorities for the castellan to surrender – which he did.

But legends had been born, legends which inspired later warriors. If a twentieth-century student named T. E. Lawrence had not decided to write a thesis on Crusader castles – which involved travelling through the Middle East – then today's world might indeed have been different.

was reinforced with projecting towers (higher than the wall) or bastions (the same height or lower than the wall). Both of these were usually round in plan, a form more difficult to undermine or dilapidate than square structures. Another impediment to tunnelling was to flood the surrounding ditch, a measure adopted in spite of the seasonal shortage of water. Walls rising from bare ground were given a sloping apron or plinth, technically known as a batter or talus. This added strength to the wall, stabilized it in the event of an earthquake, and prevented the use of scaling ladders which could not be footed upon its angled slope, and which when placed on the nearest level ground would either be too short or would be resting at such an oblique angle that they would inevitably be weakened to the point of collapse if several armed men tried to ascend simultaneously.

Another advantage of the talus was that when the defenders dropped stones upon it, they splintered and ricocheted with shrapnel effect. To enable the men on the parapets to do this without exposing themselves, stone verandahs were built out from the top of the walls with holes in the floors. Officially termed machicolation, these *meurtrières* or 'murder-holes' became a feature of all subsequent castle-construction.

Then there were portcullises and drawbridges, dead-end corridors and covered ways, so that the outer curtain walls could be reinforced or evacuated in safety. The outer, middle and inner wards could be further subdivided by walls. And very often, the allures or wall-walkways were only accessible via the doorways in the flanking towers – which meant that if an attacker did gain a foothold on the curtain wall, it did not get him anywhere. The doors into the towers on either side, though weak compared with those at ground-level,

Above: The castle which dominates the approach to the port of Sidon in Lebanon. The causeway-bridge is modern, but the thirteenth-century one was so constructed that it could be cut at an enemy's approach.
Right: Though small compared with some Welsh fortresses, Bronllys Castle's Norman tower soars impressively above its motte in marcher (border) Brecknockshire (now Powys).

Below: With such a breathtaking outlook across northern Israel, Belvoir was aptly named by the Frankish Crusaders. The modern valley reservoir is a reminder that seasonal winter rains could usually be relied upon to replenish hilltop castles' stone storage tanks.

Right: Rumeli Hisari, on the European shore of the Bosporus, just north of Istanbul. The Black Tower (left) is so called because of its evil reputation as a Byzantine prison. The other twelfth-century tower is on the right. Though remodelled by the Turks as late as 1454, they made no arrangements for emplacing the latest ordnance.

could easily withstand anything the assailant could carry up a scaling ladder – and all the time he was exposed to missiles from the inner walls as well as from the flanking towers.

For offensive action, Crusader castles were provided with a number of small, heavily fortified gateways known as sally-ports. (These are different from posterns, which are more properly simple doorways for peacetime use by foot travellers, to save opening the main gate.) From these sally-ports defenders could take advantage of the besiegers' off-guard moments, rush out, strike, and retreat hastily within the walls before the outsiders could react.

It is often assumed that the Crusaders learned all these architectural features from their opponents, the survivors returning to spread this knowledge through Christendom. But the Franks could hardly have copied the Arabs' own traditional military structures. These *alcazabas* were of rough concrete – mortar reinforced

CHATEAU GAILLARD

CHATEAU GAILLARD, on a sheer crag overlooking the River Seine near Rouen, was built by Richard I of England in territory claimed by Philip Augustus of France. Erected in 1197-98, it embodied all Richard's experiences of siege warfare during the Crusades.

The donjon was accessible only through an Inner Bailey protected by a chemise wall (like a series of overlapping bastions). This was entered over a stone bridge within the Middle Bailey, its wall guarded by one square and three round towers. This in turn was accessible only over a drawbridged ditch via the four-towered Outer Bailey, itself approached across yet another drawbridge.

But Château Gaillard had its weak points. For example, there was a privy drain under an unbarred chapel window, and the stone bridge into the chemise could afford assailants some protective cover. A combination of investment, assault and undermining throughout the winter of 1203-04 ended with Château Gaillard in French hands.

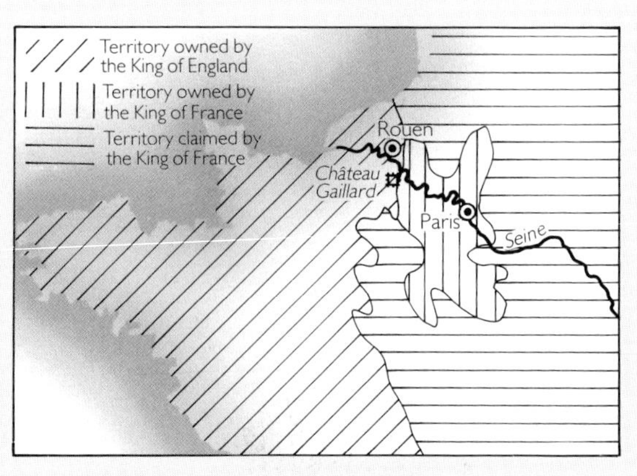

Left: Part of the fortifications of Carcassonne, rebuilt by 1285 as a royalist stronghold in Albigensian territory in southern France. (Restored by Viollet-le-Duc from 1844 onwards.)

with stones – cast between portable wooden shuttering. Inevitably, the resulting structure had to be rectangular in plan, with square towers at each corner and protecting the gate. The walls were crenellated with pointed merlons. The living-quarters were austere, and there was a mosque with a minaret from which the muezzin could call the faithful to prayer. In due course some of these *alcazabas* were enlarged and rebuilt in worked stone to become *alcazars* – defended palaces for regional military governors.

However, when the Muslim Arabs captured more settled and built-up areas, they took over and used the existing fortresses. Most of those in the Levant had been built or influenced by Byzantine engineers.

The whole of Byzantine foreign policy and diplomacy, strategy and tactics, was devoted to keeping an enemy as physically far distant, and defeat as ultimately remote in time, as possible. Military architecture was designed to make direct assault upon the walls of Constantinople (or any imperial city) so costly in lives that the enemy would be forced to abandon any siege

Above: Belver (Palma, Majorca) has airy living quarters within a thick-walled circular *castillo*. If necessary, the garrison can cross a bridge into a separate machicolated tower.

for yet another generation. Such were the fortifications developed by practical experience and by continuing study of updated ancient texts that so impressed the Crusaders that they embodied the same features in their own castles. It was these techniques that they then spread throughout Europe.

Yet if the benefits of round towers, machicolation and concentric walls were so obvious, how was it that they had not been adopted before? After all, round towers dating from Roman times had been incorporated in medieval fortifications, more commonly on the Continent of Europe, but also in the Norman kingdom of England. Indeed, several Norman castles had, from their earliest establishment, been in the form of huge round towers. These were shell-keeps, rooms for accommodation and stores being located within the hollow walls. The exterior of round towers was also cheaper to build, unlike square donjons which needed specially trimmed corner stones, all cut blocks being known as ashlar.

And what can explain the lack of early stone machicolation in Europe? Wooden verandahs called 'brattice' or 'hoarding' were customarily installed to overhang battlements and perform the same function

THE MEDIEVAL FORTRESSES OF INDIA

The sophisticated defensive structures of the Indian sub-continent were easily a match for the siege weapons and elephants of any aggressor

In the story of military architecture, as in so many other aspects of civilization, the Indian subcontinent forms a link between the traditions of the Far East and the cultures of the Middle East and Europe. Yet it is also so remote from both, and so isolated by sea and mountain, desert and jungle, that foreign influences are transmuted by the time they arrive and are themselves further altered by the Indian environment. So it is possible to trace the evolution of an Indian style of fortification produced by the climate, the terrain and the societies of the subcontinent.

Except in the alluvial floodplains – where brick can be made – stone is readily available, even in forested regions. When the rains come, they fall abundantly everywhere, producing a bountiful harvest to feed a teeming population. No matter how many die in flood, pestilence or famine, losses are soon made up. There is always a plentiful supply of warriors for army and garrison, and labourers to hew and pile up masonry in massive walls rising sheer from mountain crags. Far from water supply, such fortresses relied upon huge stone tanks or cisterns to store the monsoon rain from one season to another, distributing it to residences and animal-houses via pipes and conduits – a definite advantage over any besiegers, who had to bring all their water all the way up a steeply twisting – often stepped – roadway. Except in the immediate vicinity of the gate there was often hardly room to place scaling ladders against the wall, let alone erect a tall, top-heavy siege-tower. Even the siting of catapult artillery could not be easily done within effective range.

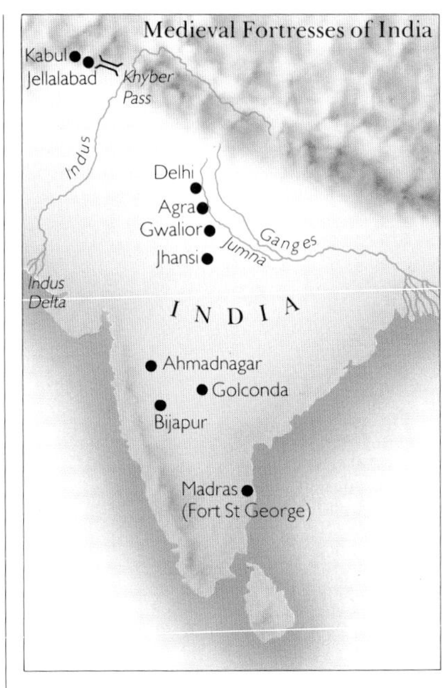

Above: The Amar Singh Gate, the Red Fort, Agra. Once through the archway across this bridge, an intruder has to turn right between the twin square and round towers, then left through the curtain wall beyond. (The British remodelled the parapets in the nineteenth century.)

The Indians did possess such equipments. There were *manjaniqs* (equivalent to *petrariae*), hurling rocks over 220 kg (500 lb) in weight; the 220-kg *maghribi* (500-lb mangonel or *ballista*); the smaller *arradah* (*catapulta*); and the crossbow *charkh* (*scorpio*). These were installed on walls and bastions for anti-personnel fire against besiegers. It may have been Alexander the Great who introduced catapult artillery to the subcontinent during his expedition of 327-325 BC.

Most assaults in India depended upon battering-rams and elephants to smash down the huge gates (often 8 m, (25 ft), high to permit ceremonial entry of howdah-equipped elephants). There were several methods of preventing a running charge straight at the doors. The flanking walls could be extended in long voluptuous curves; or the approach could be down a steep flight of steps, over a bridge, and up again; or the gateway-fortress could be fronted by a separate fortification like a ravelin; or chains could be suspended across the gateway itself. To deter elephants from maintaining sustained pressure, the solid timber of the 150-mm (6-in) thick doors was provided with iron plates and projecting spikes up to 400 mm (15 in) in length and 50 mm (2 in) in diameter.

If the besiegers broke through into the passage 3.5–5-m (12–16-ft) wide they then faced another gate, which had to be broken down in similar fashion. And now the attackers were being assailed from hidden doorways, platforms and balconies by soldiers whose armament included special tridents for dealing with the pachyderms' flanks and legs. There were also crossbeams and pillars which dropped upon the enemy when they were crowded together in disadvantageous bottleneck. Meanwhile the besieging archers were picking off the defenders, who replied in similar fashion from behind rounded or pointed merlons, or through large square or hooded loopholes cut in the merlons. From the 1350s onwards, there was also complex machicolation, often in the form of corbel-supported hooded boxes.

The customary use of anti-personnel missiles – killing people but leaving property intact – encouraged the flamboyant and intricate decoration of fortification. This added to the prestige of the ruler within, emphasizing his immeasurable wealth, his superiority to and dominion over lesser mortals. In fact, one such fortress-palace was believed to be so full of diamonds that its very name became a byword for treasure and good fortune. It is for this reason that several gold and silver mines in the United States now bear the name 'Golconda'.

The original Golconda is in the Deccan (now Andra Pradesh) in central India. Like most Indian fortresses, its rise began in about AD 1206, at which time India was a patchwork of semi-independent states. Golconda's citadel (named Bala Hissar) crowns a steep and rocky mount; 1.8 km (over a mile) of concentric stone walls run all the way around the base of the crag, and divide the strongpoint in two. Within, a sinuous writhing of lesser walls connects the natural outcrops of rock and prevents direct approach to the ruler's palace. 100 m (350 ft) below is the city of Golconda, encircled by a wall 5 km (3 miles) long, at least 12 m (40 ft) high, and 5-10 m (17-34 ft) thick. It is composed of stones (some boulder-size), roughly-hewn and bonded together, but with neater masonry around the gates. There are also 30 bastions. The outer curtain was later extended in 1724.

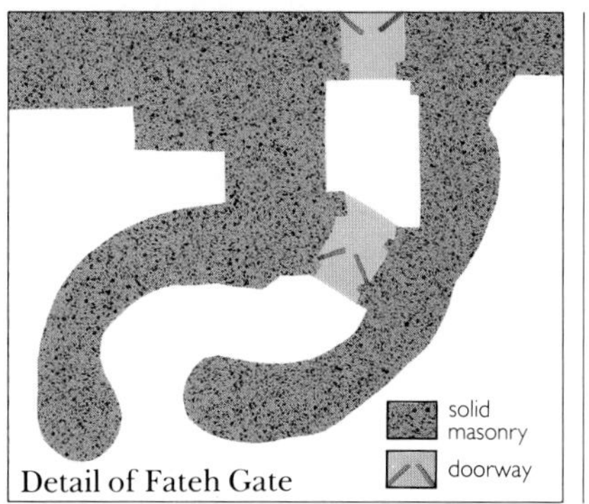

Above: Its battlemented and bastioned wall running the whole perimeter of the lofty plateau, the citadel of Gwalior looks down upon the residences below.

Left: The curving Fateh Gate into Golconda has walls 20 feet (7 metres) thick.

Below: Map of Golconda. The extension to the city wall, which includes the Nine-Lobed Bastion, was built in 1724.

Below: The Lahore Gate, Lal Qila or Red Fort in Delhi, the Mogul capital of India, was built in 1638-58.

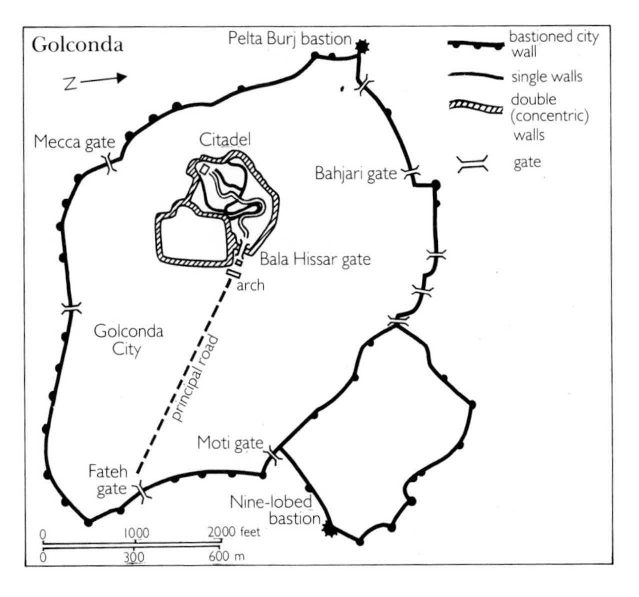

BALMORAL CASTLE

BALMORAL CASTLE, the royal residence on Deeside in Scotland, may look like a medieval fortress, but it is actually a palace built during the time of the Industrial Revolution. Prince Albert purchased the estate in 1852 and commissioned a suitably splendid and picturesque mansion. Balmoral represents the powerfully scenic results of an enduring fascination that medieval strongholds have exerted over later generations of architects and artists, authors and filmmakers.

Above: Balmoral Castle. **Below:** Its location.

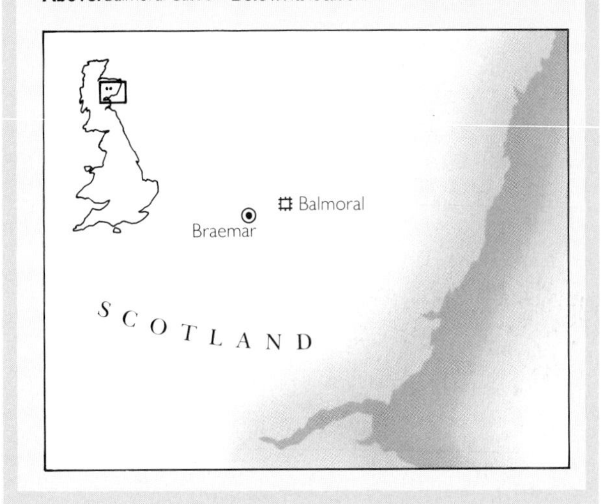

as machicolation – why not then build it in stone from the beginning? And refurbished hillforts had long demonstrated the advantages of concentric walls. Even ignoring the western European evidence around them, Frankish, Norman and Teutonic nobles visiting their kinfolk farther south or going on pilgrimage to the Holy Land, must have had plenty of opportunities to observe Byzantine and other Mediterranean fortification. So why did the widespread introduction of these architectural devices have to wait until the aftermath of the Crusades?

Part of the answer is inertia. No matter how

energetic a baron might have been, he saw no point in tearing down perfectly sound walls and towers just because foreigners did things differently. And besides, the advocates of square towers could remember seeing numbers of those in the Middle East as well as round ones. No, they would wait until a destructive siege or natural erosion so damaged their fortification that it had to be rebuilt; that was the best time for incorporating new ideas. Even then, a series of turreted concentric walls and the apparently unsupported inverted ledges of machicolation were very expensive.

The Crusades not only concentrated the minds of military men on the purposes of castle architecture, they also forced monarchs to think more carefully about finance. The raising of taxes to fund royal expeditions – and to ransom kings incompetent enough to get themselves captured – increased governmental cash flow. Inevitably some of that treasure was left behind in the coffers of the country of origin or was diverted into other projects.

Even more money became available when Orders like the Knights Templar sought ways of employing the gold and silver donated to them by the faithful. With the Crusades over, they decided to invest the money in monarchies, intending to spend the interest received in whatever charitable cause seemed best. But having dispensed that income, the kings of Europe accused the Templars of unnatural vice and witchcraft, and exterminated them – thus avoiding having to repay the loans. It was not so much the Crusades as the money they released into circulation that was responsible for the great royal castles and city-fortresses of the later Middle Ages.

THE DEVELOPMENT OF CANNON

These strongholds were to prove vital in the wars of growing nationalism, autocratic monarchy and religious dissent, the bitterness of which made dignified surrender or departure much less likely than before. European fortification had become a killing-ground, rather than a setting for brutal but chivalrous ordeal by battle.

Left: The Bastille (later a hated prison) was orginally built in the 1370s to defend the Porte Saint Antoine into Paris. Eight machicolated drum towers (with no square corners to invite dilapidation) provided fire along the face of the curtain walls, themselves as high as the towers, enabling speedy reinforcement of any threatened point.

Below: The round tower of Conisborough Castle (near Doncaster in Yorkshire) was erected by Hamelin Plantagenet (Henry II's half-brother) about 1185. It is 90 feet (27 metres) high and stands within a bailey. The buttresses seem to have been an afterthought during building; they would have been very vulnerable to dilapidation.

It was a situation in which the merchants of gunpowder and cannon found an eager market.

The introduction of these weapons from 1320 onwards did not make castles redundant overnight. If anything, it enhanced their defensive value. Early cannon, weighing up to 664 kg (1484 lb or ¾ of a ton) had to be transported on wagons and erected on timber-framed beds where they remained for most of the siege. It was much easier for the castellan to make his purchases in peacetime and to place his ordnance high up on his solid walls. The same prior advantage applied to saltpetre, sulphur and charcoal, which could be stored in comparatively dry, secure dungeons, while the besiegers had to mix their serpentine out in the

Left: The fourteenth-century gatehouse-keep of Donnington Castle, near Newbury in Berkshire, is of flint and stone. It was besieged three times during the English Civil War.

Above: King Dermot MacCarthy of Munster built Blarney Castle overlooking the River Martin in 1446. The stone conferring the gift of Irish oratory forms part of the machicolation.

open, possibly in wet or hazardous conditions. It could not be combined earlier and held ready for use because the mixture tended to sift itself into layers of constituents. The invention of corned gunpowder, in which each individual grain contained the correct proportion of ingredients, solved that problem. It also resulted in more efficient combustion, thus improving power, range and accuracy.

But the knowledge of metallurgical chemistry was still in its infancy. Guns were still liable to explode through overheating if fired more than a few times a day, or if the cannonball jammed in the barrel. It could take a long time – 12 days in the case of Constantinople in 1453 – to batter a hole in a wall. A lot could happen to the besiegers in that time, including being outranged by the altitude of the cannon on the castle walls. Certainly, gunpowder could be exploded in a tunnel under the walls, or in a petard placed against the gates with ignited fuze. (The operative had to make sure the fuze wasn't too short – or else he was 'hoist with his own petard'.)

Apart from those two devices, however, even during an actual assault, the advantages of gunpowder still lay with the defenders. All the architectural ploys of merlon and embrasure, murder-hole and portcullis, and concentric curtain walls, worked just as well for firearms as they did for hand-propelled missiles. The man up a scaling ladder was in no position to reload his piece once he had discharged it, whereas one marksman behind the battlements could keep firing as fast as his comrades reloaded and handed fresh weapons to him. If the besiegers did gain a foothold on the curtain wall, they could be swept from it by a single charge of grapeshot or langridge (small balls, nails and miscellaneous hardware) which killed and maimed, but left property intact.

Not all castles were suitable for mounting cannon. Curtain walls might be too narrow to permit safe recoil; the stoutest-roofed or most easily strengthened part of the building might not command a good field of fire; arrow-loops might have to be adapted to permit the discharge of firearms, whether pistols or long-barrelled musket-type weapons (including all those varieties of matchlock, flintlock, arquebus and so forth, of which the nomenclature can fill several pages). These latter improvements usually took the form of a round hole at the base of the arrow-slit which, if already serving both longbow and crossbow, inevitably became known as a 'cross-and-orb'. For larger artillery the arrow-loops and crenels were opened out into full-scale embrasures. Blocked with shutters when not in use, the inward part had to be splayed very wide to enable the gunners to traverse their ordnance.

Enquiring minds were meanwhile studying the

casting of stronger yet lighter barrels, and the problems of ballistics. The contribution of Charles VIII of France to siege warfare was the introduction in 1494 of wheeled artillery of lighter weight and higher muzzle velocity, with barrels that could depress and elevate, and that fired iron cannonballs. The wheels not only made for easier transportation, but also enabled gunners to traverse their pieces more smoothly and to trundle them around the siege works to take advantage of any fleeting opportunity. This mobility, plus the accuracy, enabled the artillery to concentrate on just one spot on the curtain wall, delivering an uninterrupted series of horizontal blows which eventually disintegrated the masonry. Conversely, wheels were of no particular advantage to artillerymen on the castle walls. They could still not move their ordnance from one place to another in a hurry, neither up and down stairs, nor from one tower to another.

It was becoming obvious that castle-builders were going to have to do something if they were to stay alive and in business. The first attempts involved thickening the walls of existing fortifications – a measure that soon reached the limits of practicability. Another idea was to thicken and curve the merlons on the battlements, this area being particularly vulnerable to near-horizontal fire – which was how cannonballs arrived, at the top of their trajectory if the gun had been elevated correctly. The stonework of towers and walls could also be shaped with curving or oblique faces.

But these were temporary measures. Sooner or later special castles would have to be constructed to take account of all the aspects of the development to date of gunpowder, cannon, small arms – and large standing armies.

It was the cost of this last item that really brought about the end of privately-owned fortresses and retainers, together with the additional cost of a corps of expert engineers and architects who now combined Renaissance science and theoretical mathematics with practical technology and soldiering and the stonemason's traditional skills to build the fortifications of the new age.

Left: Cornish Tintagel, the most romantic of all castles of Arthurian legend. The site had been inhabited since prehistoric times. The Great Hall (seen here) dates from 1140-75, but the castle had been abandoned by 1330.

Chapter Three

THE RENAISSANCE WORLD

The most obvious feature of the purpose-built artillery forts of the Renaissance was that compared with the lofty castles of the Middle Ages they looked low and squat. There was now no particular advantage in height. Improvements in the metallurgy of cannon and the mathematics of ballistics meant that the limitations of gravity soon came to have little effect on the performance of guns pointing upwards. In fact, too much depression was a definite disadvantage. At its most extreme, the cannonball would roll out before firing; even a slight declination would separate the ball from the propellant charge – a minute fraction of an inch would result in loss of muzzle velocity, accuracy and impact.

Conversely, an elevated cannon kept shot and powder tight together with appropriate benefits; the high-soaring ball smashed into battlements and machicolation originally built to withstand arrows and stones no tougher than their own masonry. Nor did the new artillery forts need elevated location or pinnacled tower for observation; this could be achieved by some lighter structure similar to a ship's mast. It could be replaced comparatively easily if destroyed, by which time the enemy's position would be well known. The only real advantage in altitude for smooth-bore cannon was in range, the expended shot dropping in an extended curve as its impetus faded.

THE NEW ARCHITECTURE

Another factor influencing the new forts' low profile was that the walls had to be thicker than those of earlier castles – partly to withstand the impact of hostile shot, partly to absorb the recoil and other stresses of firing. Although the first Renaissance castles were built of stone throughout, later ones had walls that were really ramparts of compacted earth and rubble held in place by solid retaining masonry or brickwork – a combination of resilience and repulse. Walls of that thickness were too heavy a mass to permit upward growth; any expansion had to be in area.

It is a mistake to think of the walls of these new forts as being tiny; they only *appeared* low in stature. They were, in fact, many times the height of a man, the whole fortress being surrounded by a deep moat (often dry)

Below: Château du Haut-Koenigsbourg, near Colmar in Alsace. Together with neighbouring Lorraine, this French territory has been frequently disputed with Germany (to whom it is known as Elsass-Lothringen). German influences predominate in this fortress, which has had gunloops cut for the new weaponry.

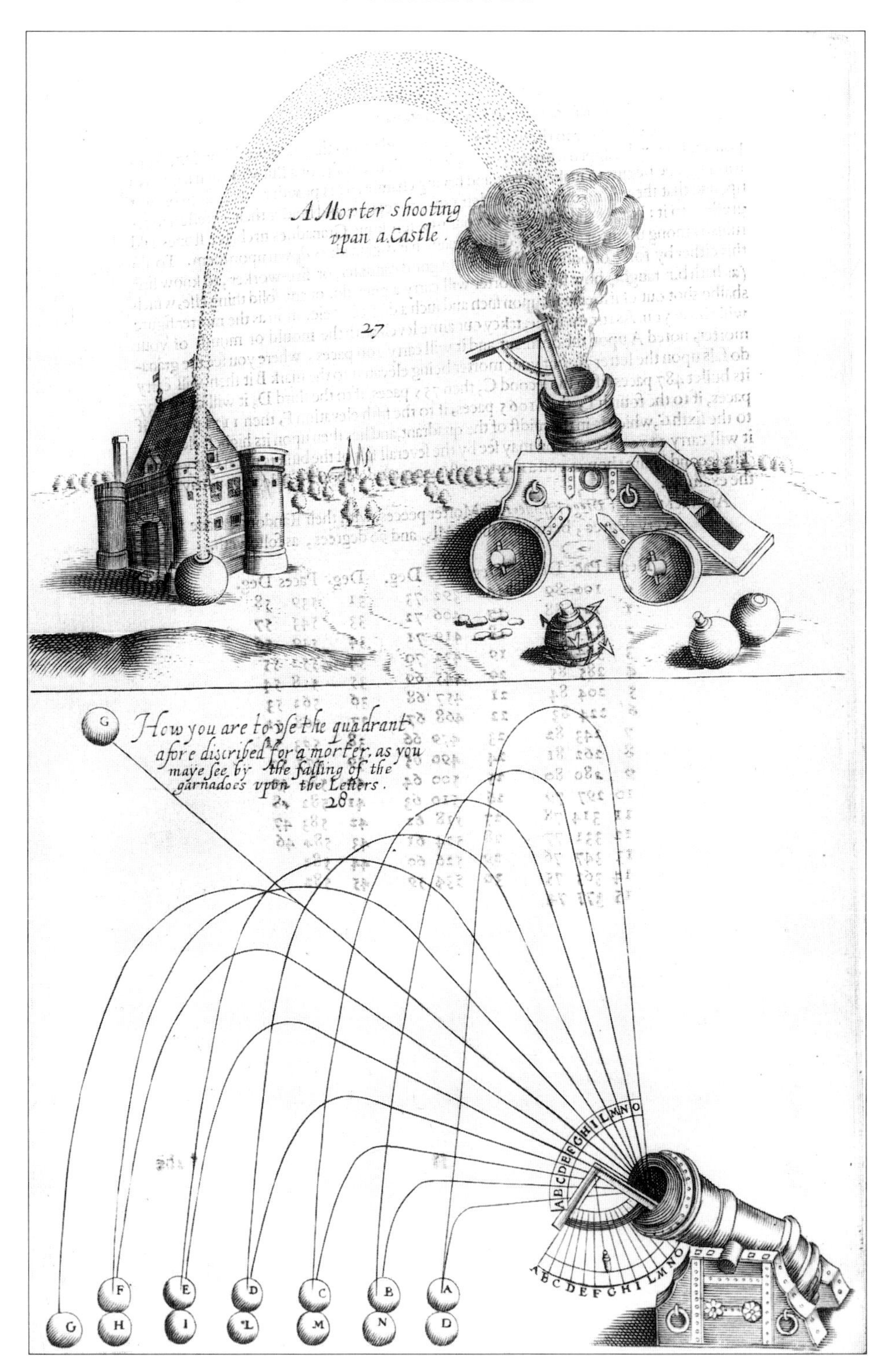

Right: The fortifications of Rhodes. The batter (talus or sloping plinth) in the foreground is all that remains of Naillac's Tower, the eastern anchorage of the chain across the entrance to the Grand Harbour. The Palace of the Grand Master of the Knights Hospitaller of St John of Jerusalem can be seen in the background, beyond the trees of the Garden of Auvergne.

Left: Post-Renaissance artillery instructions. Unlike solid-shotted, horizontally firing cannon, a mortar lobbed a hollow shell full of gunpowder up and over high walls (omitted in these pictures). A quadrant covered one-quarter of a circle. The lower diagram shows that a more vertical elevation gave a higher altitude but shorter range. Beyond 45° the range decreased because gravity influenced horizontally travelling shot. In most grenadoes (or grenades), the external fuse was lit before firing; occasionally it was hoped that the flame of discharge would ignite it (for subsequent rather than immediate detonation). The upper picture shows both smooth (for demolition) and segmented and barbed canisters (shattering with anti-personnel effect).

which prevented easy access. This obstacle was invisible at ground level. Until the attacker was standing on its rim, all he could see through the smoke was an occasional glimpse of a low wall, from whose crenels cannon belched solid ball and anti-personnel case-shot, the latter spreading a fan of death and injury like a scatter-gun.

The enemy knew that within that stronghold there were barrack-rooms, stores, reserves of food, powder and shot, all housed in safe, reinforced rooms kept dry, lit and ventilated to a standard far higher than anything in the Middle Ages. The garrison, too, had to be accommodated comfortably enough for physical fitness and mental alertness. These men were experts in the maintenance and employment of artillery. Some were clerks needing furniture and artificial illumination to cope with all the paperwork generated by the new weaponry. Moreover, the gunners could not oper-

THE KREMLIN

IVAN III became Grand Prince of Moscow in 1462. Ten years later he married Zoë Palaeologus, an imperial princess from Constantinople (captured by the Turks in 1453). With her coming arose the doctrine of the world-ruling Three Romes: one on the Tiber, one on the Bosporus, and now one on the Moskva. Ivan the Great became a Caesar (or Tsar). His capital's citadel had to have a huge wall like Byzantium, and huge round towers like Galata. Where better to find the architects than in the land of the 'first' Rome? The result was the Kremlin.

Its walls of brick (manufactured by Ridolfo Fioravanti of Bologna) form a near-triangle with a perimeter of 2.5 km (1 ½ miles). It is 20 m (65 ft) high and as much as 3.4 m (16 ft) thick. Each of the 19 towers is a self-contained fortress – but a cramped one, with room only for a couple of cannon on each level. (They were further restricted by the addition of decorated roofs in the seventeenth century.) Within the Kremlin is a complex of palaces, cathedrals and other state buildings, but in a Russian Orthodox style, not Italian.

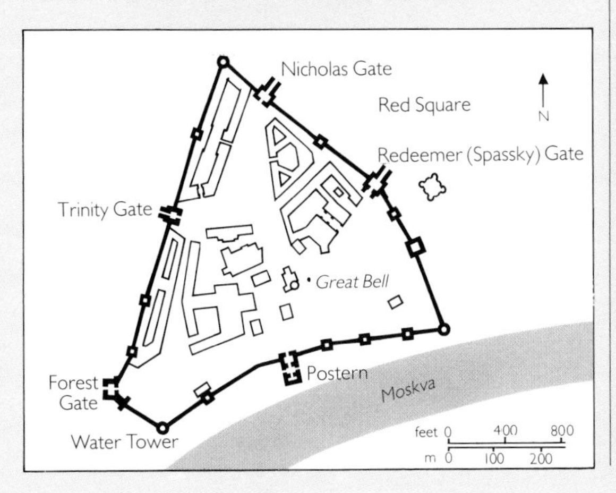

Plan of the Kremlin.

Red Square: the Kremlin's Spassky Gate (once the Gate of the Redeemer).

ate their pieces effectively if nearly asphyxiated in badly ventilated casemates.

When a medieval castle was fully manned, it accommodated hundreds of knights, men-at-arms and archers, each one wielding a single sword, pike or bow. One man – a score of men – more or less would not make a great deal of difference. But the artillery forts of the Renaissance relied on small teams operating a few cannon of efficient mass destruction; one guncrew disabled through illness, injury or malnutrition could mean the loss of the fort. In addition, new approaches in understanding disease and its prevention were naturally reflected in the design of living-quarters.

And so was the new attitude towards the importance of the individual brought about by the Reformation. The ordinary soldier might not have appreciated it, but the State was beginning to assume responsibility for equipping, training, feeding, clothing, paying and entertaining him – even if the last-mentioned activity only involved compulsory Church parades on Sundays, and preventing drunkenness by flogging him if he over-indulged. No baron ever bothered to supply his

Right: Cross-and-orb gunloops at St Mawes Castle, opposite Falmouth in Cornwall.
Below The three levels at St Mawes seen from the south. In wartime, the eighteenth-century naval cannon would not be left on such exposed display.

levy of peasants with weapons, let alone anything else. The artillery forts of the Renaissance and succeeding centuries took account of all these new trends.

Most of all, they represented genuinely *military* architecture; fire bases, infantry depots and signalling stations, manned by full-time professional soldiers. If no war threatened, the garrison was usually stood down to care-and-maintenance status; nobody else lived there. They did not serve as royal or baronial palaces. Monarchs now lived in totally civilian complexes such as Hampton Court or Versailles, far removed from their frontier forces and fortresses, yet at the centre of a network of communications maintained by despatch

THE CAMPING OF THE KING AT MORGVISON

Left: Henry VIII's siege-train in 1544, when the English Army (denoted by the Cross of St George) captured Boulogne. There are arquebusiers, pikemen, halberdiers, cavalry and a variety of cannon. Casks are standard stores containers, one wagon laden with liquid refreshment. Cooked food is purchased bottom left. Nearby, a unit is briefed. The conical carts seem to be for transporting tents. Beyond, the enemy city burns, while violent rain and wind compounds war's discomforts. Hats and tents are blown away, including one accommodating ladies attached to the army. A frightened horse refuses to move and one man, depressed by it all, buries his face in his hands.
Right: Twentieth-century military architecture atop sixteenth-century Fort St Elmo, Malta.

riders on fast horses, and surrounded by a new class of technical experts, advisers and managers of finance, religion, agriculture, navigation, and all the other affairs of state. Warfare was now just one of many departments to be delegated to trusted subordinates. Only the most eccentric or truly professional king actually led his troops in battle.

Nevertheless, most princes still took more interest in their armies than in any other branch of government. And in this sphere, as in any other, they could – according to their personality – be influenced by a flattering or competent advocate of some particular weapon, tactic or campaign during the Renaissance period of new ideas and technology. In terms of military architecture, the development of English coastal fortresses under the Tudors illustrates the search for the most effective type of fortification.

HENRY VIII: DEFENDER OF THE REALM

With plenty of money to spend after his seizure of monastic property, Henry VIII wanted a strategic weapons system. Artillery forts were to be built to dominate coastal shores where a French army might invade, especially the approaches to vulnerable naval bases. Thus delayed and funnelled into disadvantageous killing-grounds, their troops would be defeated by the English army. Meanwhile the Royal Navy would be dealing with the assault ships (already disabled by the coastal forts) and the covering task force – and would then go on to carry the war into the Continent.

The early programme construction was based on the ideas of the Bohemian Stephan von Haschenperg. His castles were round towers, surrounded by round bastions, which enabled cannon to be moved around to bear on the enemy through splayed embrasures and crenels. In some instances the plan-view of these castles looked like a Tudor rose or a trefoil; in others the central keep was encircled by one continuous battery or had two wings of fortification. But whatever the design, its most notable feature was that it was curvilinear. On paper it was pleasing to the eye and in theory it was a perfect defence, not a single flat surface being presented to cannonballs. Even the merlons were rounded vertically, to deflect shot up and over the heads of the gunners on the roof.

In practice, however, it had one serious defect: it was very difficult for defenders to fire at the dead ground in front of adjacent bastions – unless the embrasures were

TUDOR COASTAL DEFENSE

The epitome of the classic 'forts-fleet-field army' strategy of Henry VIII kept Britain's shores safe from foreign invaders

In 1538 Henry VIII instigated the construction of completely new forts at Sandown, Deal, Walmer and Sandgate in Kent, at Calshot and Hurst on the Solent (Hampshire), and at Portland in Dorset. Blockhouses (simpler structures mounting just one cannon with a couple of musket embrasures) were erected: two at Tilbury and three at Gravesend on the Thames, two at Cowes on the Isle of Wight, and one at Sandsfoot opposite Portland. Three new batteries were added to existing fortifications at Dover, and an old tower at Camber in Sussex was expanded into a full-sized fort – all by 1540. Blockhouses were then constructed on Brownsea Island in Poole Harbour and at Harwich in Essex. St Mawes and Pendennis Castles denied access to Falmouth in Cornwall, and two blockhouses (Dale and Angle) guarded the entrance to Milford Haven. The Humber was protected by a fort and two blockhouses at Kingston-upon-Hull.

Above left: Henry VIII, King of England 1509-47.

Above: 1520: French and English flags wave and cannon roar a salute from new blockhouses as Royal Navy warships take their sovereign to meet Francis I on the Field of the Cloth of Gold. In the background, the Norman castle of Dover looks down upon the joyous proceedings – an interlude in near-continuous hostilities with France.

Left: Perspective section through Deal Castle.

Deal Castle exemplified the round, perfectly symmetrical form advocated by the Bohemian Stephan von Haschenperg. The circular central tower or keep was surrounded by six semicircular bastions of slightly less elevation. It was then ringed by yet another lower and wider chemise of six semicircular bastions. Each floor and each roof mounted guns with overlapping arcs of fire through embrasures in the walls or between the merlons. The whole fortress was enclosed within an artificial ditch, covered by musket fire from ports cut in the lowest level of the outer bastions. The only entrance was over a portcullised drawbridge into one of the outer bastions. Once there, attackers would have to batter their way into the central keep before gaining access to other parts of the fortress.

Not all of Henry's forts were as complex as Deal. Some lacked the artificial ditch, relying on the lowest level of cannon-embrasures to sweep the ground of approaching infantry. Walmer and Sandown (Kent) were completed with simple central towers surrounded by a separate complex of four semicircular bastions. St Mawes had a trefoil chemise of three bastions; Portland dispensed with separate bastions – two wings joined the keep to a curved battery pierced by five embrasures. Pendennis was composed of a keep completely encircled by one continuous battery. Hurst's tower was not round but 12-sided (with three bastions). Sandgate's chemise of bastions formed what might best be described as a three-pointed ellipse. Camber had five semicircular bastions, the connecting curtain walls forming shallow Vs with their apexes towards the enemy. This was taken a stage further at Hull, where there was a square castle with two ogival bastions on opposite sides of the structure.

Above: Deal Castle was not besieged until 1648 (during the English Civil War). The Royalists held out against the Parliamentarians for two months, but surrendered after the total defeat of a seaborne relief force.

Below: South-coast castles and other fortifications built or reconstructed during Henry VIII's fortress-programme.

Henry VIII was moving away from von Haschenperg's 'round' ideas towards the mainstream of Italian fortress design. This is evident at Southsea Castle, begun in 1543. It had a square keep, surrounded by a bailey, also square but at right-angles to the keep. The north and south curtain walls thus formed angled bastions in the Italianate fashion. The east and west corners were formed into rectangular gun platforms composed of solid earth and capable of mounting the heaviest cannon. A dry ditch surrounded the entire complex.

Henry VIII's fleet-forts-field army strategy worked. In spite of the loss of his great ship the *Mary Rose* during the attack on Portsmouth in 1545, the French fleet still could not force its way through the Royal Navy and past its supporting forts. They did not attempt to land troops in the face of Henry's army, but drew off and contented themselves with harrying the Isle of Wight before sailing away.

To forestall the use of the island as a permanent base, Henry VIII ordered two more castles to be built there. The one at Sandown had a V-shaped angled salient; the other at Yarmouth was given an arrowhead bastion. Subsequent additions to existing fortifications had similar bastions, usually made of earth to absorb the smashing impact of cannonballs. Impressive facades of brick or stone were simply thin retaining walls.

The English development of arrowhead bastions was continued in the last year of the reign of Mary Tudor (1558) and completed under Elizabeth I. These fortifications were located at Berwick-on-Tweed, the English bridgehead on the Scottish bank of the border-river. The whole town was ringed by a polygonal wall, the arrowhead bastions projecting from the corners. The cannon mounted on them covered all the approaches to the walls, and without storming them no army could reach the bridge over the Tweed on the road to the south.

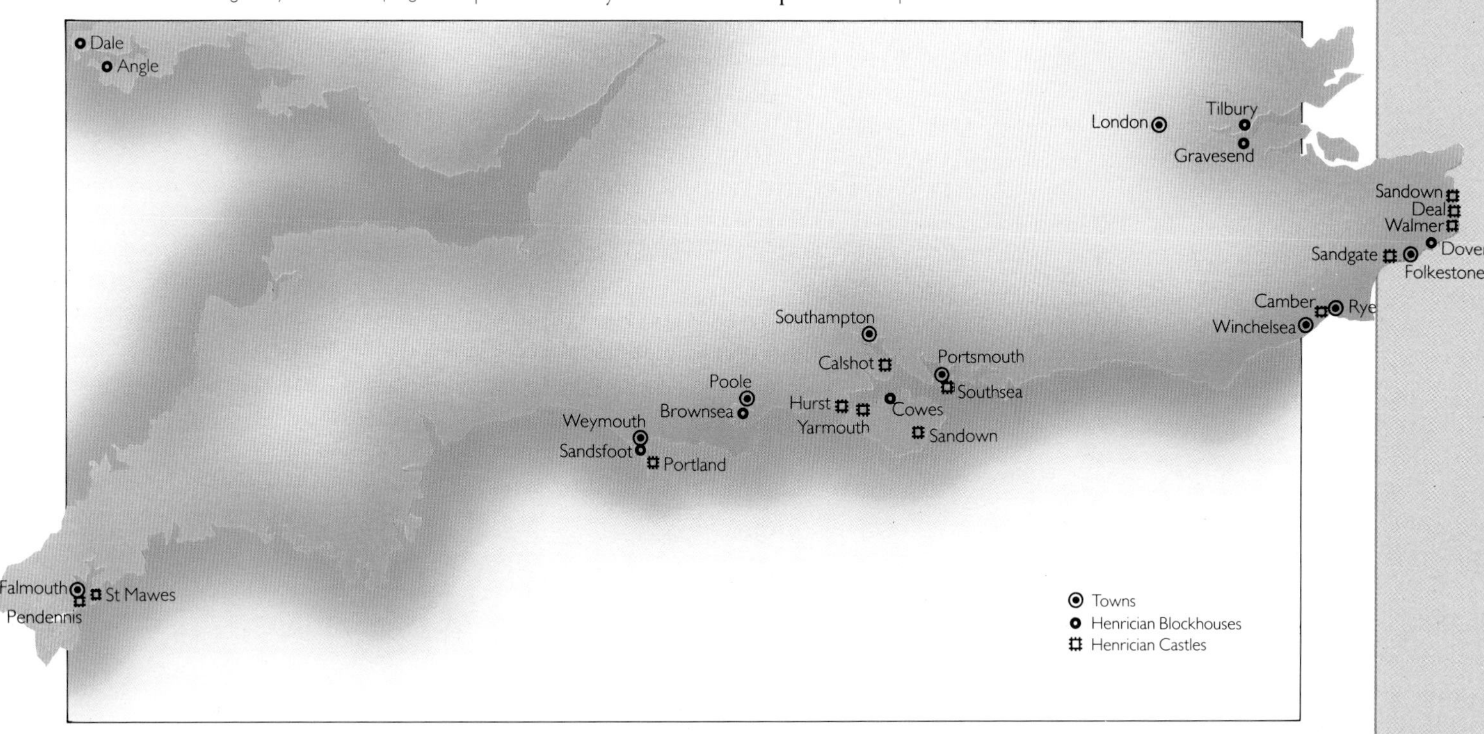

Left: Tilbury Fort: the Water Gate (erected in the reign of Charles II). In 1897 (when this informal photograph was taken), it was the headquarters of the Thames District Establishment, still guarding the seaward approach to London.

widened so much that the gunners themselves became exposed. For these coastal forts were intended not only to exchange cannonballs with a fleet offshore, they had to repel traditional investment, to beat off an infantry assault while the land-facing guns broke up siege-works and smashed enemy artillery. Every minute the fort held out enabled the seaward guns to keep firing at the main body of the enemy's fleet, preventing it from landing the principal invasion force, thus gaining time for Henry VIII's field army and warships to come into action.

However, as the construction programme developed momentum, so the design of fortresses changed. Whether this was because the deficiencies of these forts became so apparent that Tudor architects suggested and implemented improvements – under royal supervision – or whether more information became available concerning Italian military architecture is not known. What is certain is that from the beginning of the construction of Deal Castle in 1538 to the completion of Yarmouth Castle in 1547 there was a complete shift from German to Italian ideas.

Italian ideas did not entirely eschew curved surfaces. Some bastions were ogival in plan-view – like a pointed arch with the apex towards the likeliest direction of hostile approach to deflect oncoming shot. While practical experience suggested this development, Renaissance disciplines of geometry and trigono-

HAARLEM

A Netherlands town that in the sixteenth century was a symbol of Dutch and Protestant rebellion against Spanish and Catholic tyranny. At that time its walls were only 6 m (20 ft) high with round towers and a ravelin in front of the northern Kruispoort. It contained a population of 18,000 non-combatants and a garrison of 6,000.

11 December, 1572: Don Federico de Toledo arrives outside with 17,000 soldiers.

18 December, 1572: Spanish bombardment begins.

20-21 December, 1572: Spanish assault on Kruispoort repulsed. Spanish begin digging trenches towards the Kruispoort ravelin.

15 January, 1573: Haarlem garrison abandon the Kruispoort position. Spanish enter the town to find themselves facing a *retirata* (an internal ditch and bastioned rampart) formed by demolished houses.

31 January, 1573: Spanish assault repulsed. Coal-miners from Liege in Spanish Netherlands begin tunnelling under the walls, and are intercepted by Haarlem counter-miners.

25 March, 1573: Haarlem garrison sortie captures Spanish food convoy.

28 May, 1573: 100 Spanish vessels under Count Bossu defeat Martin Brand's 150 Dutch ships on Haarlem Lake. No more food for Haarlem.

9 July, 1573: Spanish cavalry defeat Dutch army.

12-13 July, 1573: Haarlem surrenders.

14 July, 1573: Grand entry of Spanish into Haarlem. They have lost 12,000 men. Dutch losses are 4,200 (plus citizens dead of starvation and disease). Another 1,735 of the garrison and citizens are executed by beheading or drowning.

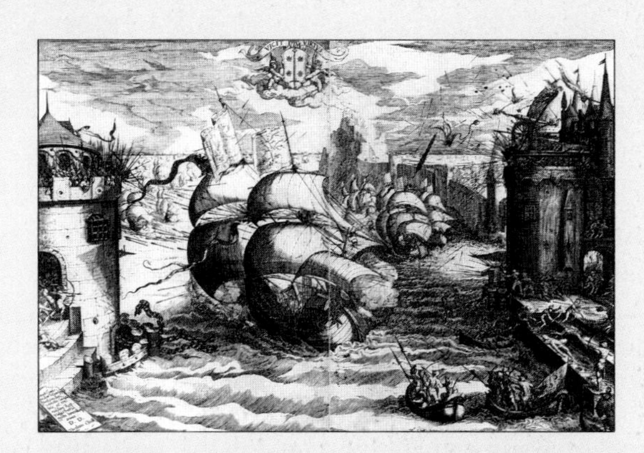

Above: A 1595 illustration of Haarlem: dramatic rather than accurate.

Below: Whoever controlled the Zuider Zee could blockade or relieve Haarlem.

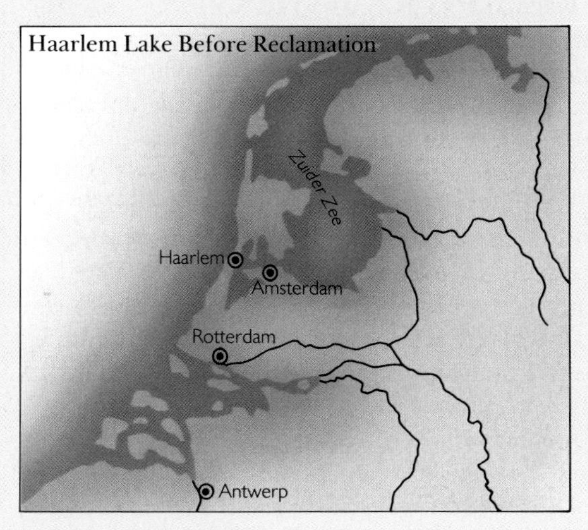

metry demonstrated mathematically that bastions with a V-shaped ground-plan were the most effective defence. With guns mounted in the base of the V, no assault party could approach the wall without being fired at from the side. And if those particular guns were sited in a recess, they would themselves be screened from direct fire. This resulted in a bastion of arrowhead plan which, with variations, became the most typical feature of European fortification for the next three centuries.

THE RENAISSANCE MASTERS

The names of Albrecht Dürer and Leonardo da Vinci are often associated with the history of military design. Dürer, born in Nuremberg in 1471, wrote, illustrated and printed a treatise on fortification that was not really the first of the genre, but was the first to popularize the subject. However, it had little or no influence on practical construction. The designs Dürer proposed were merely of medieval round towers, updated to mount and ventilate cannon in casemates, and with vertically-curved merlons to deflect shot. It can be argued that some German cities owe their remodelled defences to Dürer's ideas; it may also be imagined that Stephan von Haschenperg had a copy of Dürer's book in his luggage when he arrived at the court of Henry VIII – but both of these connections remain conjectural.

Leonardo da Vinci (1452-1519), however, is universally regarded as a scientific genius – a fortress-architect and far-sighted inventor of siege machinery and other weaponry. And yet there is no evidence that his patrons ever hired him either to design any fortification or to manufacture any novel military hardware. In this field he was a complete theorist, with little knowledge of the practical problems of heat, chemistry or mechanics. Indeed his private miniaturized jottings and sketches even suggest a physical short-sightedness which might have prevented his personal observation of fortresses in the landscape.

Two other Renaissance artists did nevertheless achieve much more in the form of solid military architecture. Michelangelo (1475-1564) was in Florence at the time of its rebellion against the Medici. The republican government appointed him Commissar-General of Fortifications. He organized the construction of bastions and a moat, and fortified the hill of San Miniato, based on the plan of the 1496 works at Ferrara 130 km (80 miles) to the north. They were really fieldworks – earth ramparts reinforced with crude bricks, lengths of timber and bundles of sticks (fascines). This form of construction absorbed shot, and was also cheap and easy to throw up. The sculptor-artist must have done his work well. The enemy bombarded the fortifications but did not attempt to storm them. Instead they stopped all supplies getting into Florence.

Left: The Siege of Malta, 23 June 1565. The Turks push forward from batteries and siegeworks bisecting the narrow neck of Mount Sceberras. Sheer weight of numbers overwhelms the Knights of St John in the star-shaped Fort St Elmo. Now Turkish galleys ringing the island can enter the Grand Harbour for direct assault upon the walls of the main Christian stronghold on the twin peninsulas opposite. Meanwhile, the Turkish field army will attack the landward defences of Senglea and Birgu. But Mustapha Pasha is not optimistic as he considers the 4,000 dead it has cost him to gain Fort St Elmo: "If the son, which is so small, has given us so much trouble, what will the father do, which is so large?"
Right: Tour du Marinka, Corsica: a typical Mediterranean headland tower, lookout and refuge when pirates threatened. The entrance door was usually halfway up the tower, only accessible – for at least part of the approach – via a removable ladder. The initial flight of stone steps seen here appears to be of more recent construction. Note the handgun loops in the parapet, and the fringing machicolation for vertical missiles upon immediate assailants. Note too the dense, fragrant, undergrowth; in World War II, this gave its name to the resistance workers who sheltered in it – the Maquis.

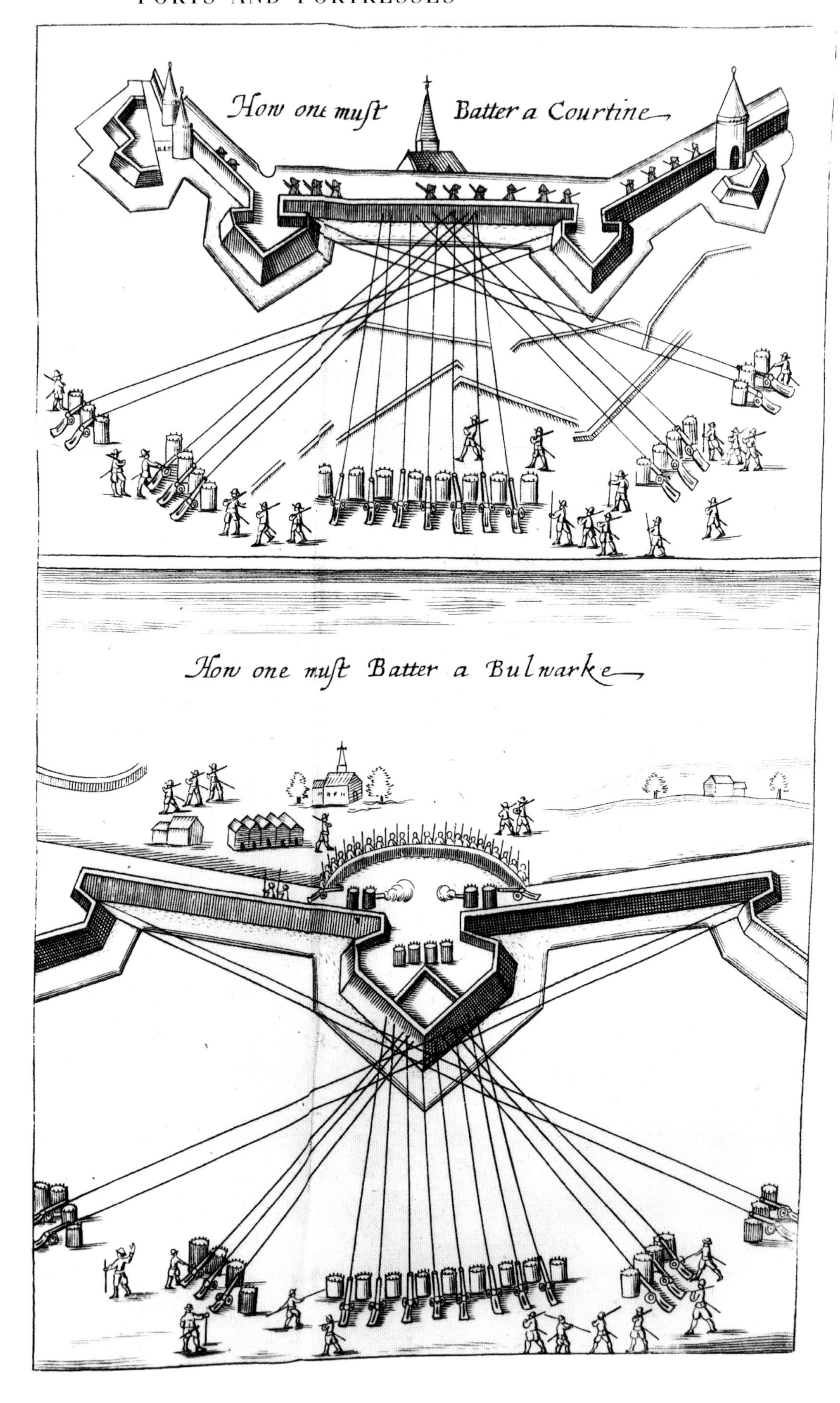
How one must Batter a Courtine
How one must Batter a Bulwarke

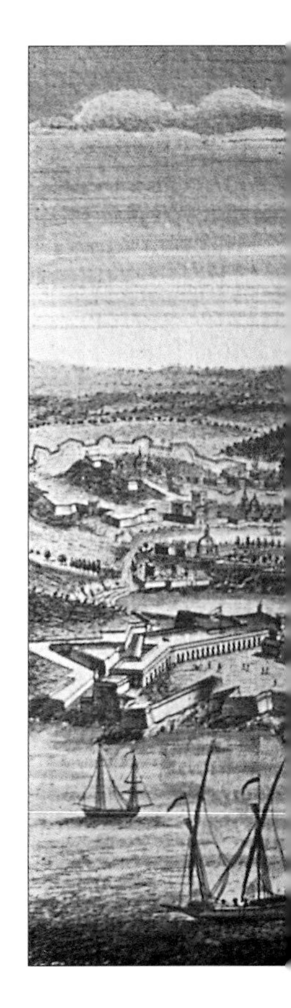

Far left: 1673 instructions for artillery concentration against a curtain wall (**top**) and an arrowhead bastion (**bottom**). Other fortifications include hornwork and ravelin fronting left and right gates, respectively. Within the lower bastion is a permanent retirata, with an improvised one cutting off the whole bastion, should the besiegers succeed in storming it. Both sides make use of gabions, while two sentries parading in front of the guns show that (a) the cannon are firing indirectly at full elevation and (b) they are protected from enemy fire by their own fieldworks.

Left: Eighteenth-century bird's-eye view of Valletta from the sea.

It took more than nine months of investment before the city was starved into submission on 2 August, 1530.

Another, earlier, artist with both Florentine and military connections was Filippo Brunelleschi (1377-1446). His chief claim to fame is Florence Cathedral, in particular its great dome, but of even more significance was his employment – if not outright invention – of a huge revolving crane. Made of wood and counter-balanced with blocks of stone, it enabled masonry – and the heaviest cannon – to be raised to the highest levels and there positioned accurately. It was a great improvement on the timber trestles and earth ramps previously used to lift ungainly weights.

Brunelleschi remained essentially a civilian artist and architect, who undertook military commissions only incidentally. It is very likely, however, that military engineering has always been a specialist discipline. The Crusader-scientist Petrus Peregrinus of Maricourt is one whose name has come down to us; another is Simon the Dyker, at Ardres in France in about 1200. There must have been many more. Nevertheless, it is during the sixteenth century that the building of fortresses begins to pass totally out of the hands of general builders and designers and away from gifted amateurs. This era saw the rise of the full-time specialists in the construction of fortification, men who had no other title or rank or place in society.

Particular among them is the Florentine dynasty of the Sangallos, associates of Michelangelo. Francesco Giamberti was born in 1405. He had two sons (Giuliano di Francesco Giamberti and Antonio da Sangallo the elder) and one daughter, Smeralda. Her son was Antonio da Sangallo the younger. They were not only architects but designers and constructors of building plant, contractors and quantity surveyors, consultants, accountants and instructors. Their works included the castle at Rome's port of Ostia, a three-sided fortress with two drum towers and a polygonal keep, its apex pointing towards the approach of greatest peril. In 1487, the Sangallos provided the town walls and citadel of Poggio Imperiale with ten bastions. At Nettuno (1501-1503) on the coast south of Rome, the four triangular bastions were recessed where they joined the corners of the fort. Technically called 'retired flanks', this produced an arrowhead plan, the first certain examples of this type of structure. What looked like the 'barbs' were termed 'orillons', and could be either rounded or acute-angled.

One of the Sangallo students was Micheli di Sanmicheli, born in Verona in 1484. Returning from Florence to Venetian employment, he began rebuilding the walls of his native city in 1530. Naturally he designed bastions, but he revetted their earth ramparts with bricks. The walls were further strengthened with inner buttresses (or counter-forts) arched to form a protected corridor running below the parapeted wall-walk. In plan-view the bastions had straight sides joining them to the walls. However, the upper level of

TURKISH HIGH TIDE

From the East the Turks came to capture Christendom but the spirited defence of Crete, Malta and Vienna turned the tide

In 1310, the Order of the Knights Hospitaller of Saint John of Jerusalem established their headquarters on the island of Rhodes in the eastern Mediterranean, just 10 km (about 8 miles) from Muslim Asia Minor. An outpost of Christian civilization, the Turks regarded it as a lair of pirates who murdered and robbed Saracen and Arab traders, enslaving the survivors in infidel galleys. By 1480 Sultan Muhammad II (the Conqueror of Constantinople) had had enough. He despatched an amphibious task force of 160 ships under Meshid Pasha, arriving off Rhodes on 23 May.

At this time, the city, including the Grand Harbour waterfront, was completely surrounded by walls, each sector defended by men (and women) from France, Germany, the Auvergne, Spain, England, Provence and Italy. There were outer ramparts and a ditch, and each gateway was a fortress. So too was the citadel (or Collachium) which included the Knights' Hospital, their Auberge (or barracks), and the Palace of the Grand Master. A chain was stretched across the mouth of Grand Harbour from St Angelo's Tower on Windmill Promontory to Naillac's Tower. This stood at the base of another north-running promontory, on the end of which was the fortress Tower of St Nicholas, separating the Grand and Ancient Harbours. The shallow parts of the latter had spiked planks fixed to the seabed.

An initial Turkish rush against the West Wall was halted. They then bombarded the St Nicholas Tower prior to an amphibious assault. Their galleys were specially suitable

Below: 12 September 1683: Sobieski attacks the huge Turkish camp outside Vienna, just as the Ottomans prepare to storm the city.

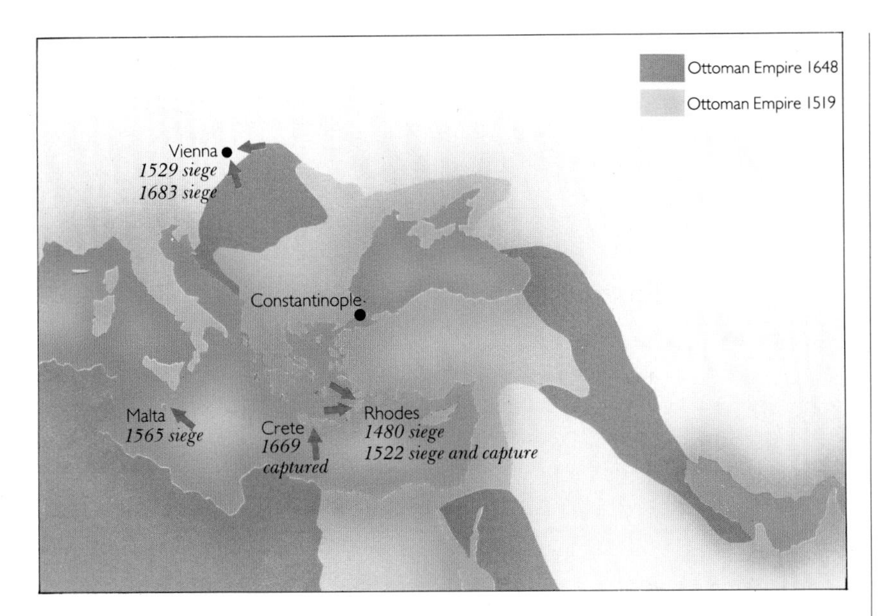

The map **above** shows the expansion of the Ottomans between 1519 (when they held Egypt, Palestine, Asia Minor, the Balkans and the northern Black Sea) and 1648, having added North Africa, part of Arabia, Mesopotamia, and Hungary.

Above: The Turks' floating-bridge breaks adrift and disintegrates, preventing their assault on St Nicholas Tower, Rhodes, 1480.

for this, being of shallow draught and having their guns mounted in the bow. They were repelled. The Turks reorganized and concentrated on the southeast Wall of the city. But when breaches were made, Meshid Pasha saw that the Rhodians had already improvised interior defences. Not only that, the Knights broke out, smashed the siege-works and retired safely. So Meshid Pasha ordered another assault on the St Nicholas Tower via a floating bridge. But an Englishman named Rogers swam underwater and released the anchor. And when the Turks tried to get their bridge into position by means of boats, it was wrecked by Rhodian cannon.

The Turks suffered such losses in two consecutive assaults on St Nicholas that they did not try again. Instead Meshid Pasha organized artillery bombardment all round the city, while concentrating for another attack on the southeast corner. This time their sappers tunnelled into the dry ditch and then began to erect an earth ramp that would eventually reach the height of the city walls. It was at this time that the Knights' trebuchet came into its own.

This type of weapon had been introduced in about AD 1147 and was still in use despite the invention of gunpowder. It relied for propellant-power not upon the elasticity of sinews and twisted hair but upon a solid weight of several tonnes attached to the short arm of a huge pivoted beam. The missile was slung from the end of the long arm which

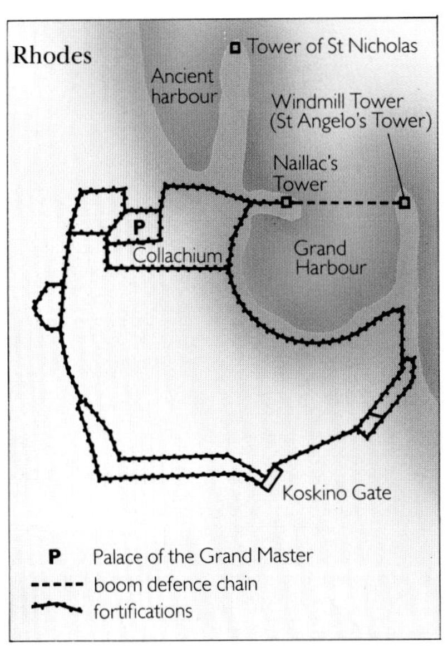

The map of Rhodes **above**, shows how the St Nicholas Tower dominated the approaches to both the Ancient and Grand Harbours. The latter was further defended by the Naillac and Windmill Towers, with a huge chain suspended between them.

was winched down to the ground and then released. Its range was about 450 m (about 500 yd) but its greatest advantages were its very high trajectory and its accuracy; the counterweight could be slid up and down the arm to vary the range. On Rhodes in 1480, its huge stone missiles not only killed the ramp-builders but penetrated the ground and collapsed the mineworkings. A sortie completed the destruction.

With no sign of surrender (in spite of the stakes of impalement displayed around the city) Meshid Pasha ordered further artillery bombardment, culminating in another assault during which part of the southeast wall collapsed. This time the invaders included a team of assassins with orders to kill Aubusson, Grand Master of the Order. He was indeed wounded but the assailants were driven out and the whole Turkish army started to retreat to their ships. As they sailed away on 18 August, 1480, two Spanish ships arrived with supplies and reinforcements.

In 1522 the Turks were back, under the personal command of

Above: Malta, 28 June 1565: having captured Fort St Elmo, the Turks concentrate against Senglea and Birgu (with Fort St Angelo at its tip). Note floating bridge and boom-defence chains.

Left: A splendid portrait of Suleiman I – the Magnificent – born 1494, Sultan of Turkey 1520-1566.

Sultan Suleiman the Magnificent. By now the Knights of Rhodes had not only repaired the damage done in the 1480 siege and the 1481 earthquake, they had also strengthened the works. The entrance to the Koskino Gate, for example, was over a drawbridged ditch, on to a round artillery bastion, turn left, then right, over another drawbridge into an Italianate triangular bastion, turn left, bear left, turn right, and through the Inner Gate in a wall 12 m (40 ft) thick. Villiers de l'Isle-Adam's garrison comprised 700 knights and 5,300 serving men.

The first of an estimated 100,000 Turks came ashore on 28 July, 1522. Mines and counter-mines were a feature of this siege – after the Turks had spent August filling in ditches to get close to the walls. They mounted a total of seven massive assaults before what should have been an overwhelming offensive in November 1522 – which was repulsed. By now Suleiman had lost 60,000 men – killed or died of disease – and he had other campaigns to fight in addition to this. Moreover, winter was coming on. Accordingly, he proposed an armistice. An honourable surrender was arranged, and on 21 December, 1522, 180 Knights of Saint John and 1,500 serving men left Rhodes.

On the mainland of Europe, Suleiman's power was expanding irresistibly through the Balkans. On 27 September, 1529 the Sultan stood with 125,000 troops before the walls of Vienna, in particular before the weak medieval southern sector. They dug mines and made several attempts to storm the Habsburg capital. The wall was breached, but inside the Count of Salm and his 16,000 defenders had improvised a *retirata*, a fieldwork that still prevented entry to the city. As in Rhodes before, Suleiman decided it was too late in the season for lengthy campaigning, and withdrew his army on 14 October, 1529.

Meanwhile the Knights of Saint John had been looking for a new base. In 1530, they accepted the offer of Malta from the Holy Roman Emperor Charles V. (It had been part of his Spanish domain.) The

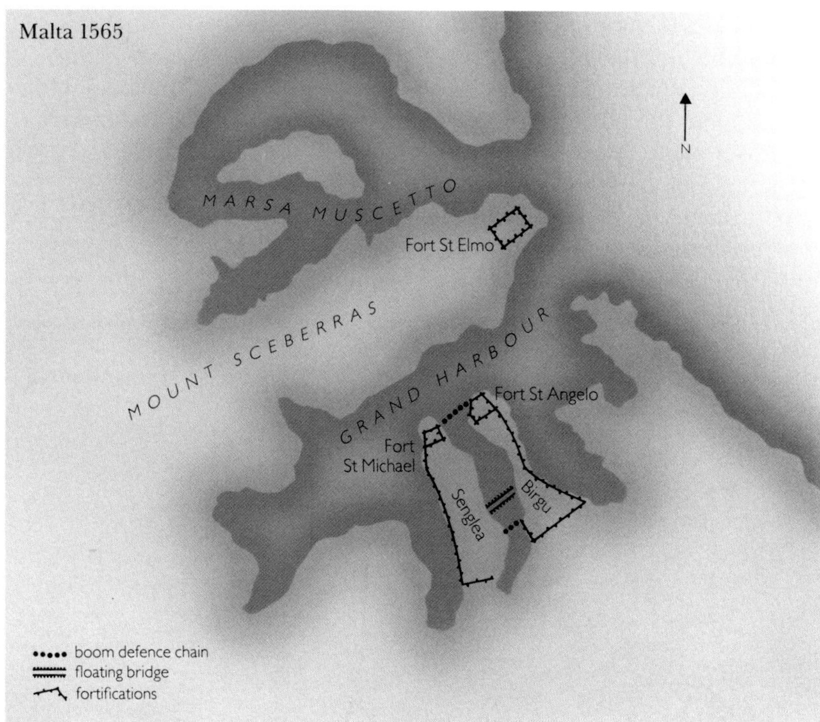

Above: After 1565, Valletta was built on Mount Sceberras. In 1798 Napolean ejected the Knights. The Maltese rebelled, helped by the British. The Royal Navy became the dominant factor on Malta and, with Army, RAF and Maltese forces, defended the island against sea and air raids and blockade during World War II.

Knights had little to do with the Maltese but established their seven-galley fleet base on a complex inlet on the north of the island. The narrow peninsula of Mount Sceberras divided it into two basins, the eastern one becoming known as Grand Harbour. Two east-west peninsulas there (Senglea and Birgu) were fortified according to the latest ideas by Bartolomeo Genga, becoming Fort St Michael and Fort St Angelo. A chain prevented waterborne access between them, and they were designed as one fortress. A lesser work on the end of Sceberras was called Fort St Elmo.

On 18-19 May, 1565, with orders from Suleiman the Magnificent to exterminate the Christian corsairs once and for all, Mustapha Pasha and Piale Pasha (the Turkish general and admiral) started landing an eventual total of 36,000 men on Malta. They decided to capture St Elmo first. With that in their hands they could use the western basin of the harbour and make shipborne attacks on the two main fortresses. That objective should have been attained in six days. Fort St Elmo was eventually captured on 23 June, 1565: 1,300 Christians had been killed and 4,000 Turks, including Dragut Reis, the Sultan's naval commander-in-chief who had arrived to take charge. 600 Christian reinforcements also landed in Malta and made their way through enemy lines and by boat to Birgu.

However, the Turks proceeded with a close investment of the twin fortress by land and water. Their assaults seemed never-ending, although a cavalry attack from the Maltese capital at Medina did bring temporary respite. The Turks then returned to the attack with even greater vigour. But at last on 7 September, 1565 a full-sized relief force arrived from Sicily. At first Mustapha Pasha ordered the remnants of his army to embark, but when he learned that the reinforcements numbered only 6,000 fresh troops he started to land his weary soldiers again. In this confused atmosphere they met the Christians and were defeated. On 11 September, 1565 they sailed away, leaving 30,000 Muslim and 8,000 Christian dead.

These defeats can now be seen as the high tide of Turkish expansionism in Europe. It did not seem like that to the people who lived on the borders of the Turkish Empire for several further generations. In 1683 (under Sultan Muhammad IV) the Turks again besieged Vienna. By now the city had been refortified according to the ideas of the German architect Daniel Spreckle (1546-1589) who could not help being influenced by Italian designers. The Turkish general was Kara Mustapha Pasha. His 138,000 troops followed their usual pattern of siege warfare: they dug closer and closer to the walls, getting ready for that great assault when they would push forward through any breach regardless of loss. But they had not prepared adequate defences for themselves. On 12 September, 1683 King John Sobieski arrived from Poland with 30,000 troops. Cooperating with the Austrians, he fell upon the Turkish positions. By nightfall the Turkish high tide was indeed receding.

their flanks was set back, and the bottom level had embrasures to enable fire to sweep the ground – in effect an early version of a caponier. Sanmicheli also set the fashion for strong yet decorated entrances to fortresses when he designed Verona's Porta Nuova.

Sanmicheli not only designed fortresses in Italy, he also fortified Venetian colonies overseas, assisted by his nephew Gian Girolamo Sanmicheli, the latter's brother-in-law Luigi Brugnoli, and the up-and-coming Savorgnano dynasty of military architects. Between them, they (and in turn their students and local site managers) built and updated Candia, Canea and Retimo in Crete; Famagusta and Nicosia in Cyprus; and Sebenico, Zara, Zante and Corfu in the Adriatic.

TURKISH EXPANSION

These places were not so much bases of operations as fortress-refuges for Venetian expatriates: traders,

Roman Catholic priests, administrators, their families and their servants, who – stiffened by small full-time garrisons – hoped in emergencies to hold out until an amphibious task force could arrive from Venice to put down local Greek Orthodox rebels or to drive out Muslim invasion. For at this period it was not yet the Pacific and Atlantic that dominated world trade but the Mediterranean Sea – the very area in which Turkish expansion was relentless and ruthless.

Spinalongha, north-east Crete: a Venetian outpost-fortress built 1579; unoccupied by the Turks until 1715. When the latter abandoned it in 1904, it became a leper colony. It is now an empty ruin.

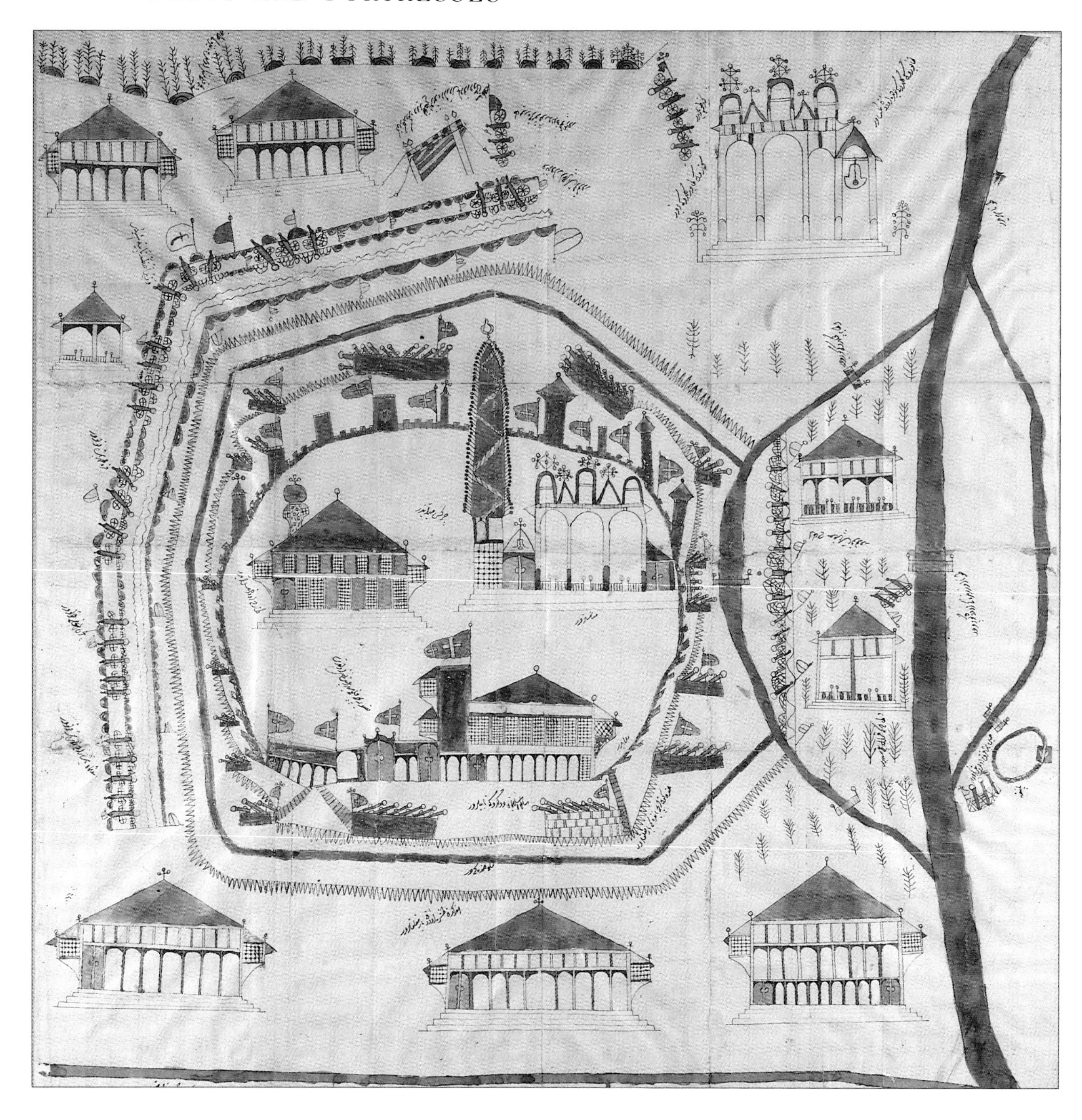

Turkish sketch of Vienna, 1683. Note Danube and St Stephen's Cathedral. There is an inner bastioned, turreted and battlemented wall; a series of entrenchments, batteries and improvised redoubts; and an outer rampart. The Turkish lines comprise: circumvallation; main infantry position; and ditch and rampart fronting wheeled artillery. There is no defence against attack by a relieving army.

70,000 Turks fresh from the storming of Nicosia on 8 September 1570 arrived outside Famagusta ten days later. Blockade and steady attrition weakened Marcantonio Bragadino and his 7,000 Venetians and Cypriots until the main assaults began in April 1571. These took the form of trenches dug as deep as the Famagusta ditch and right up to that obstacle itself. The besiegers then made a covered double embankment to the very rampart, which they could thus begin to undermine or scale. The process cost them 30,000 men, but the remainder still outnumbered the garrison by six to one and many of the latter had themselves become casualties. With no hope of relief, at the beginning of August 1571 Bragadino arranged an honourable surrender and, he hoped, eventual repatriation for his men. The Venetians and Cypriots marched out in ceremonial style on 6 August, 1571, whereupon Mustapha Pasha ordered them to be seized and enslaved. But first they had to repair the Famagusta fortifications.

The fear of Turkish atrocities inspired the construction of fortified refuges throughout the Christian Mediterranean. The western basin was a Genoan sphere of influence. Genoese forts represented a theme of architecture dating back to the Roman watchtowers on the German frontier (the *burgi*), the Celtic *brochs* of Scotland, and the lookout towers erected by the ancient Hebrews in the fields of Canaan. They served as sentry posts in time of war or when pirate raids by sea or land might be anticipated. Bonfires ignited on their roofs at the approach of corsairs gave the locals time to run and hide in the hills.

Some towers were big enough to serve as refuges for the villagers and their portable valuables. Lightly-armed bandits might not be able to undermine the walls or bring with them a siege-tower capable of reaching the entrance half-way up the smooth wall, nor cannon capable of battering through 4.5 m (15 ft) of masonry. Many of these towers had been built in ancient times and kept in some sort of functional repair ever since.

The Genoese had a policy of erecting them wherever they went. Early ones were square; later ones (including the huge Galata Tower at Constantinople) were round. The ones constructed under Genoese direction in the sixteenth century were also round, about 12 m (40 ft) high and 13 m (45 ft) in diameter, had a parapet fringed with embrasures and loopholes, and looked very much like the castle (rook) on a chessboard.

One of them was on a headland overlooking San Fiorenzo Bay in northern Corsica, a headland cloaked with myrtle-bushes and hence named Mortella Point.

Meanwhile other Italian architects were hiring out their expertise to the rulers of northern Europe. In 1544 Girolamo Pennachi showed Henry VIII how to set out a pentagonally bastioned camp for the siege of Boulogne, works extended the following year when the English were themselves blockaded. Then in 1558 Giovanni Portarini and Jacopo Contio advised Sir Richard Lee, an Officer of Works, on how to deal with the problems that had arisen during the fortification of Berwick-on-Tweed.

King John III of Sweden (ruled 1568-1592) invited four Milanese brothers named (presumably in Swedish) Paar to build Borgholm Castle and Kalmar. Also from Milan Pietro Antonio Solario and Alevisio da Milano travelled to Russia to build the Kremlin in Moscow in 1485-1499.

Simone Genga (1530-96, and of the Florentine dynasty of architects of that name) accompanied Stephen Báthory from Transylvania in the Balkans on the latter's election to the Polish throne. The Italian advised the Poles on how to build and on how to conduct siege operations, in particular against Pskov, the fortress-city obstructing passage round the southern end of Lake Peipus on the borders of present-day Estonia. This winter campaign lasted from 18 August, 1581 until 15 January, 1582, when the war ended in a stalemated armistice.

There was also Francesco Paciotto d'Urbino (1504-76). Variously known by his middle or his last name, he was employed by the Duke of Savoy. His speciality

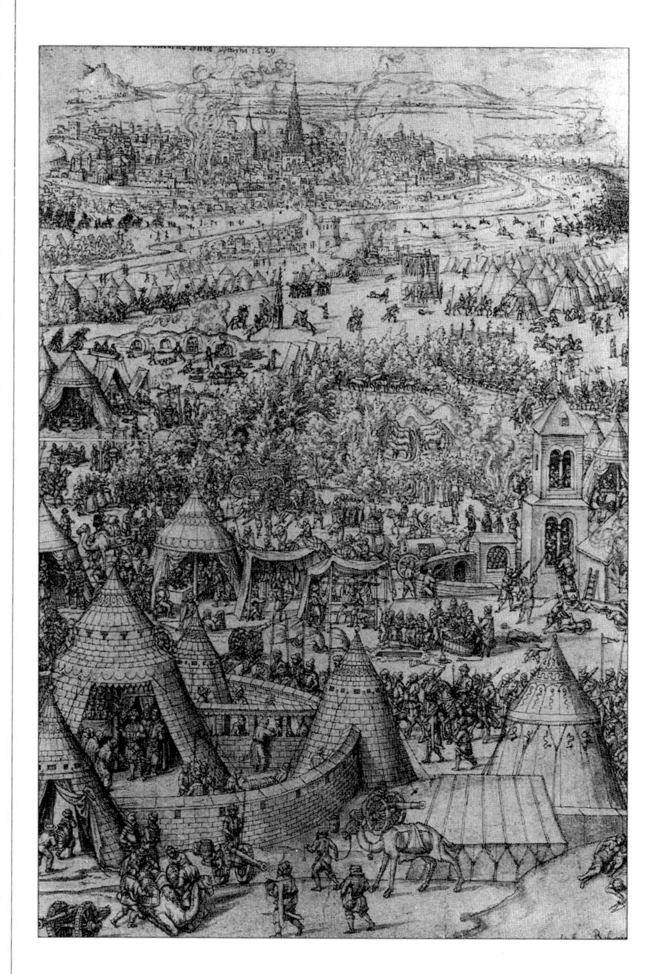

Last days of the Turkish siege of Vienna, 1529. In his stylized tented camp, Suleiman the Magnificent considers the scarcity of supplies and decides to abandon the siege. Christian prisoners' pleas are ignored; they are massacred, some thrown into the flames of burning churches and plunder too heavy for loading on camels; beautiful women are enslaved.

VAUBAN

The technical mastery of Vauban dominated the science of fortification for a hundred years

Vauban introduced the systematic application of 'sap-and-parallel' trenches. If a circumvallation was vital, then it was dug. Usually, however, Vauban ordered only as much as was necessary. Narrow trenches were pushed out towards the walls in a series of short zigzags or 'traverses' to prevent a single cannonball from hurtling right along the trench, even though its mean heading was straight towards the enemy. The front sapper was protected by a wheeled mantlet, while overhead cover was provided by a wooden gallery. Gabions (baskets full of earth) raised the height of parapet and parados along both the approach and the lateral trenches which were pushed out parallel to the original entrenchments. The front-line troops then moved forward to the new positions.

As much work as possible was done under cover of darkness and when it had been repeated often enough, Vauban's troops were close enough to the enemy for a causeway to be filled in across the moat or for a sap to be driven under the walls and a gunpowder mine installed.

Simultaneously Vauban was coordinating the fire of carefully sited batteries so that – during daylight, at any rate – they, not the infantry, prevented a sortie by the garrison. Meanwhile other cannon concentrated on the weakest section of the wall. Vauban also deliberately employed ricochet fire to eliminate troops sheltering behind ramparts or in the covered way before they had to be assaulted by infantry. He did not encourage the bombardment of the town itself; the object was its capture, not its destruction. At the appointed time, the cannonade reached a climax of bombardment and covering fire, the walls crumbled, the gunpowder mine was detonated, the assault troops stormed the breach and captured the fortress. That was how Vauban captured Maastricht after just 13 days of siege in 1673; similar success attended other operations of his.

Left: Vauban.

Top right: Pictorial explanation of all the terms used in fortification and siegecraft, as regularized in Vauban's time. Note the *cheval-de-Frise* or turnpike. In a later generation, such obstacles topped tollgates to prevent reckless horsemen leaping over without stopping to pay; hence the roads themselves became known as turnpikes.

Bottom right: Belgian Namur (or Namen) is not only a manufacturing town, it also straddles the confluence of the Sambre and Meuse valleys providing various routes between France and Germany. In 1695, it was captured from the French by William III of England and the Netherlands.

To ensure that his own techniques did not rebound against their inventor, Vauban used the peacetime interludes between wars to rebuild French fortresses in such a way that they could withstand – or delay even longer – the type of siege operations he himself practised. Studies in geometry and mathematics established the best form these new fortifications should take before construction began.

Vauban was always ready to adapt his ideas according to geography, but to him the ideal fortress was an eight-pointed star, some 900 m (1,000 yd) in diameter. The walls joining the bastions ran in apparently straight lines – 'apparently', because each face was recessed for the middle third of its length, thus enabling it to be covered by fire from the flanking and projecting other two thirds, themselves covered by *their* flanking bastions. Ideally, there were four gates at approximately the four points of the compass, their approach roads making several changes of direction over the outworks – fatally confusing for attackers, but hardly inconveniencing the march of friendly troops.

The outworks formed a total of 16 points, one fronting each wall and each bastion, all surrounded by a complicated series of moats, all able to fire into the flanks or rear of infiltrators, and all provided with overlapping artillery arcs to keep the besiegers' cannon at too far a distance to damage the central fortress. Attention was paid to the solid-

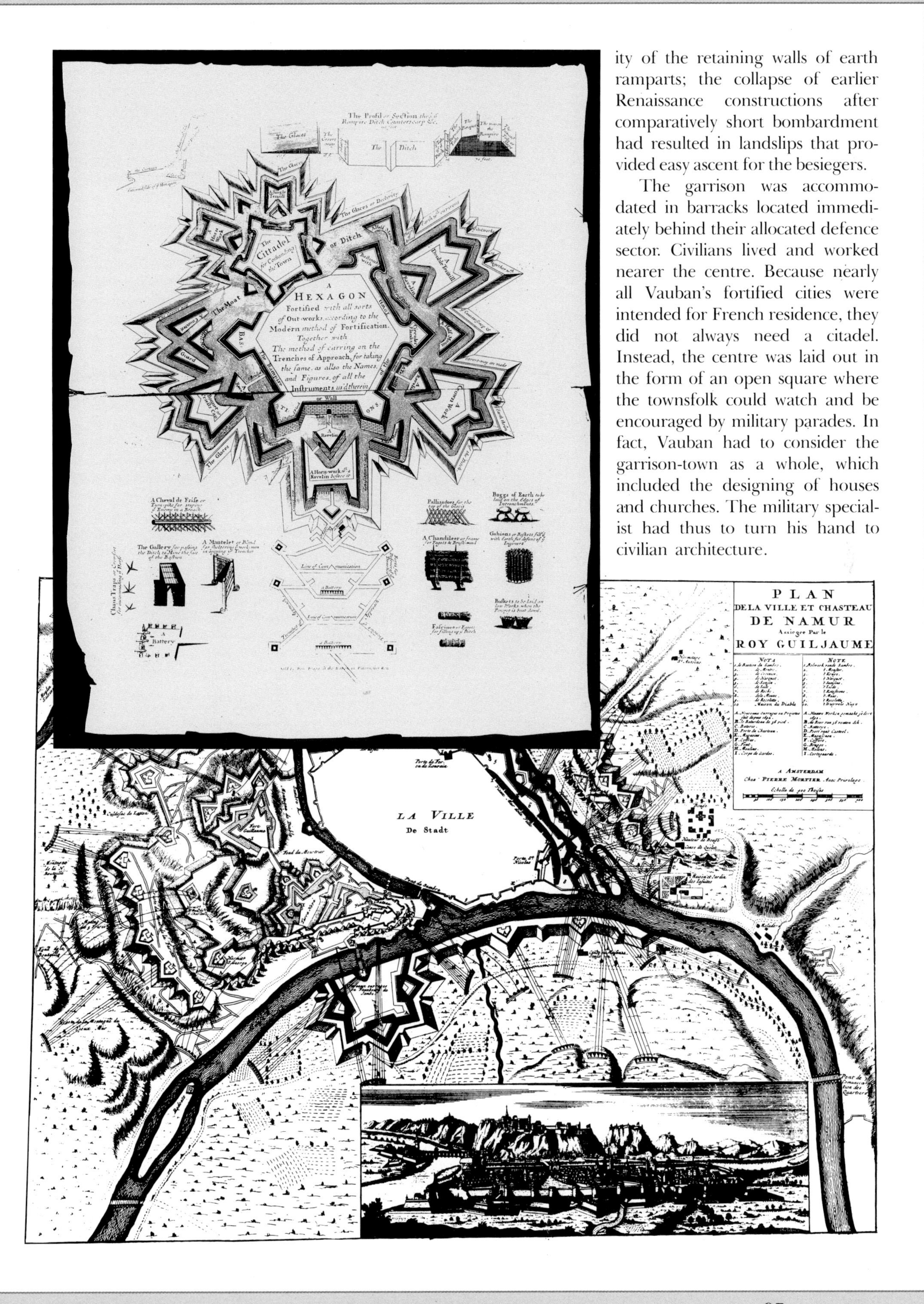

ity of the retaining walls of earth ramparts; the collapse of earlier Renaissance constructions after comparatively short bombardment had resulted in landslips that provided easy ascent for the besiegers.

The garrison was accommodated in barracks located immediately behind their allocated defence sector. Civilians lived and worked nearer the centre. Because nearly all Vauban's fortified cities were intended for French residence, they did not always need a citadel. Instead, the centre was laid out in the form of an open square where the townsfolk could watch and be encouraged by military parades. In fact, Vauban had to consider the garrison-town as a whole, which included the designing of houses and churches. The military specialist had thus to turn his hand to civilian architecture.

was the citadel, a fortress within a fortified town in a hostile land. The city had to be held as a base for field army operations while simultaneously denying strategic communications and industry to the enemy. However, the citizens (being residents of that country) could be relied upon to collaborate only for as long as fear or self-interest prevailed. If that failed, they might at any time make common cause with their fellow-nationals outside, and turn upon the foreign garrison within their midst. So the garrison needed a strong refuge into which it could retire, from which it could dominate the town, and from which it could speedily emerge into the countryside beyond.

The citadel proposed by d'Urbino and his disciples as the ideal was a pentagon, its corners extended into narrow-throated arrowheaded bastions, the whole surrounded by a ditch which traced a five-pointed star. The far side was cleared into a glacis killing-ground. The only access was via a bridge and a gateway-fortress through the straight wall opposite the apex of the pentagon. A fortified command-post stood in the centre of the citadel. Ideally, the citadel itself stood in the centre of the fortified city, the latter's streets radiating from it, so that the garrison's grapeshot could clear the town of rioters or invaders. If geography forced the citadel to be located at one end of a city, then the city's streets should be laid out on a gridiron pattern for similar tactical cannonade.

Of course it was rare for geography to permit perfect orbital symmetry. A low, flat plain was best for

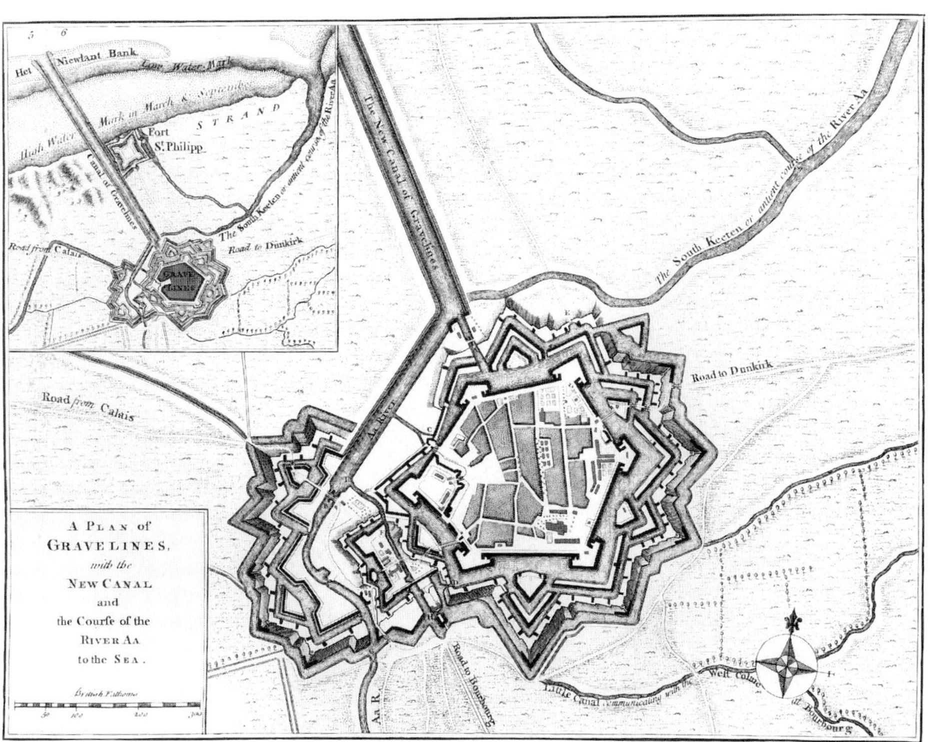

Above: 1761 plan and location map of Gravelines, a coastal fortress-town in the disputed territory between France and the Netherlands.

Left: The fortified city of Valletta, built after the 1565 Turkish siege of Malta. Fort St Elmo is on the northern tip of Mount Sceberras on the left. During World War II, these ramparts mounted anti-aircraft and coast-defence guns.

that – the sort of terrain found predominantly in the Netherlands. And in 1567 that country was also an area in which occupying troops could not rely upon the loyalty of the residents in the garrison towns, much less the surrounding countryside. And so it was, that d'Urbino and his team were invited there by the Duke of Alva, Regent of the Spanish Netherlands. The Antwerp Citadel was built in four months.

DUTCH DEVELOPMENTS

Considering the importance of water in the traditions of the Netherlands, it is not surprising that the Dutch made use of wet moats around their fortresses, of which the enceintes (enclosed sites) were at one stage obviously derived from d'Urbino's ideas. But by the time of Prince Maurice of Orange-Nassau (1567-1625), the Dutch were teaching other countries about fortification.

Prince Maurice himself captured 98 fortresses by storm or siege, and had relieved 12 others, so he can be regarded as something of an expert on the subject. He was also keen to pass on this knowledge to subsequent generations of Dutchmen and their allies. In collaboration with his former tutor, Quartermaster-General Simon Stevin, he drew up textbooks and established a school devoted to fortification and surveying at Leyden University.

In practical construction, the Dutch employed ravelins – free-standing, near-triangular quadrilaterals – one in front of each of the pentagon's 230-m (250 yd) faces. The earliest ravelin had been built at Sarzanello, between Leghorn (Livorno) and Genoa, in 1377. Then there were demi-lunes (half-moons, or lunettes), similar to ravelins but with a concave rear face, which fronted each arrowhead bastion. A horn-work was a squarer, larger and more complicated version. And so in due course was the bonnet, or priest's cap. These could be sited beyond the main moat (although perhaps surrounded by an extension) to provide fire over a wider area of approach. A crown-work was similar but was bastioned, while tenailles and double-tenailles were smaller versions. Slightest of all were the counter-guards: long, near-triangular positions located wherever there was need for extra covering fire within the moat.

Access to all these positions was via drawbridges or tunnels. The latter enabled strongpoints to be held,

reinforced, or evacuated even after they had been surrounded. This was particularly important in relation to the counterscarp (whose loopholes in the far retaining wall of the ditch provided fire across the moat into the rear of the attackers) and in relation to the covered way, an infantry trench running all the way round the full length of the outer perimeter.

The counterscarp gallery was the devising of one of Prince Maurice's successors as an expert in fortification, Baron Menno van Coehoorn (1641-1704). Another Dutchman, Sir Bernard de Gomme (1620-1685), crossed over to serve Charles II in England. His principal achievements (the fortifications of Portsmouth and the construction of Tilbury Fort) have all the classic Dutch characteristics.

The actual work of building these fortifications was undertaken either by local labour (forced or paid) or by available serving soldiers. Not until late in the seventeenth century were regiments of sappers and miners established, although artillerymen were among the earliest to form their own units.

It has, however, become evident that long before then there were specialist *officers* concentrating on the design and construction of permanent fortresses, temporary fieldworks, and the devices and techniques required to disable both. In fact, the single word 'engineer' became synonymous with the design and creation of military works; when, generations later, other specialists began designing and building canals, tunnels, roads, railways and dams, the word 'civil' had to prefix their claim to be 'engineers'.

Above: 1390: an Anglo-French amphibious force assaults Medina in Ifrikiya (modern Tunis), retaliation for attacks by Barbary corsairs. Crossbows and stone-throwing bombards have had little effect so far. The garrison clusters behind raised drawbridges, waiting to make a sortie.

Right: In an attempt to revive his Swedish empire of the Baltic, collapsing through Russian victory and separatist movement, Charles XII besieged the Norwegian fortress of Frederikshald. On 12 December 1718, he was struck and killed by a cannonball.

Military architects and engineers benefited not only from practical army service and disciplined training but also from the Renaissance custom of studying past history and the handbooks and critiques of other nations' devices. The armies of France, Spain and Germany were thus no less behind developments – as far as circumstances allowed – and contributing their own novelties and variations.

The French returned from their late fifteenth and early sixteenth century campaigns in Italy with both expertise and experts. In operations against the English on the Channel coast, against the Spanish in the Pyrenees and the Netherlands, against the Germans along the Rhine frontier, and against the Huguenots within their own borders, the French learned how to take fortresses and how to build them. By 1598, Henri IV was king of a united country. His Superintendent of Finances was the Duc de Sully. He also controlled the departments of government dealing with artillery, building and fortifications, bringing to them an accountant's efficiency. One of his engineers was Errard de Bar-le-Duc (1554-1610), French traditions and knowledge being subsequently handed on through the generations of the Chevalier de Ville (1596-1656), Blaise François, Comte de Pagan (1604-1665) and the Chevalier de Clerville (1610-1677).

De Clerville became Commissar-General for Fortifications and Town Reconstruction. Although he did not actually command armies in the field, he conducted a score of successful sieges and would have become more famous in history had it not been for the greater renown of his protégé, Sébastien le Prestre de Vauban.

VAUBAN AND HIS SUCCESSORS

Vauban was born in 1633 and joined the French cavalry at the age of 18. At that time the minority government of Louis XIV, controlled by Cardinal Mazarin, was dealing with a rebellion of disaffected French noblemen supported by Spanish troops. Much of the campaigning took place in the territory bordering Alsace-Lorraine and what is now Belgium (but was then the Spanish Netherlands). While still a cadet, Vauban saw service on the walls of Clermont-en-Argonne and (under de Clerville) in the siege-works of Sainte-Menehoud, both fortresses dominating the road from Germany to Paris where it passes through the hilly region west of Verdun. By the time he was 22, Vauban had commanded the engineers repairing the captured fortress of Sainte-Menehoud, and the forces besieging four other towns (including Clermont-en-Argonne, which had changed hands). He was appointed Engineer-in-Ordinary to the king in 1655.

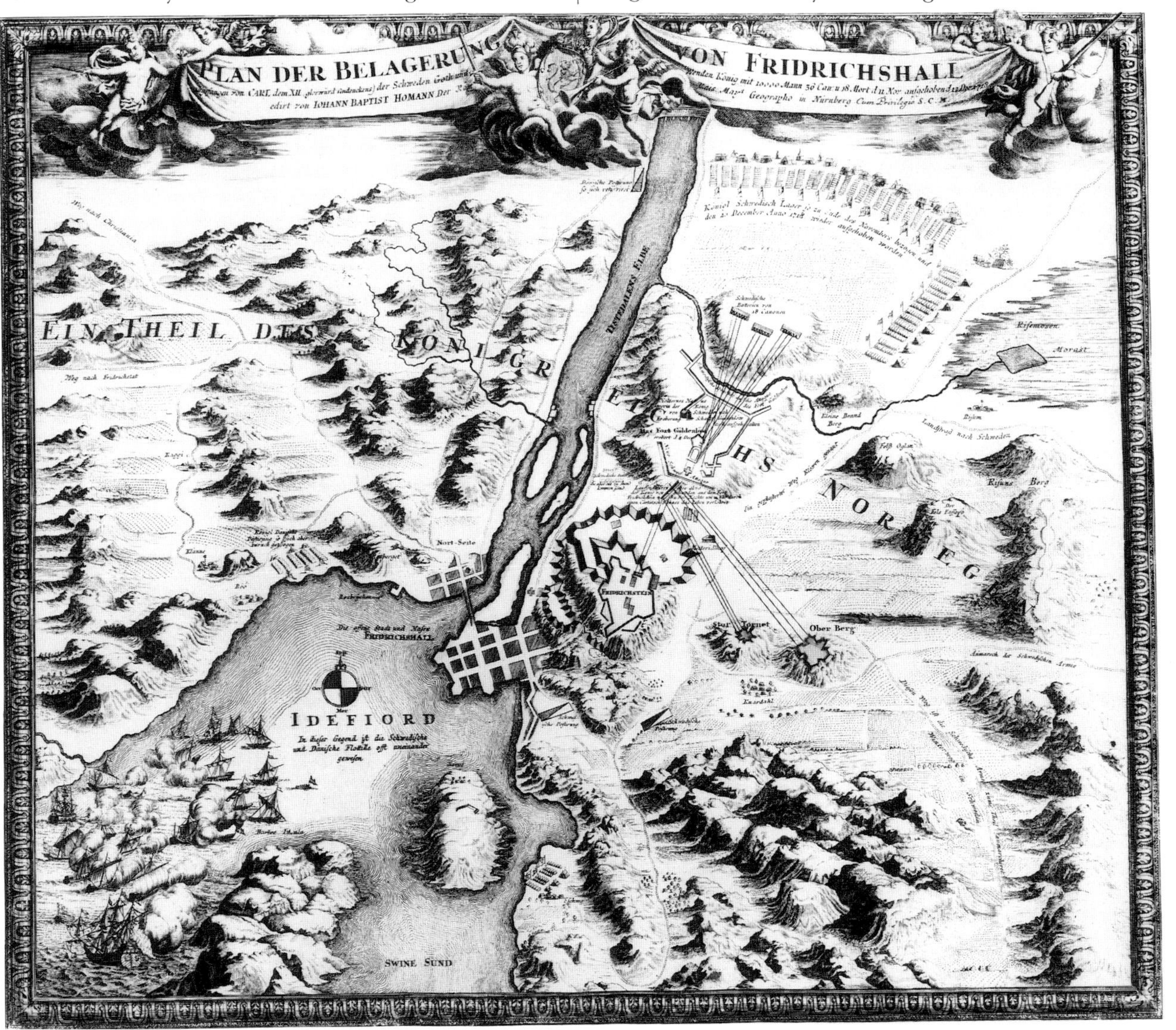

MARTELLO TOWERS

MARTELLO TOWERS derive their name (in a transmuted form) from a sixteenth-century Genoese fortlet on Mortella Point in Corsica. In February 1794 it took three days of naval bombardment and infantry assault before the 38-man garrison surrendered. The stubbornness of their defence prompted the British government to erect similar 'Sea-fortresses' in Guernsey, South Africa, Canada and Minorca. The main programme, however, was instigated by the danger of a Napoleonic invasion of Britain in 1803. Between that date and 1814 a total of 164 were built, mainly on the southeast coast of England, but also in Ireland, Jersey and the Orkneys.

An average Martello looked like an inverted flowerpot, 10 m (32 ft) high and 19 m (62 ft) in diameter, tapering to 12 m (49 ft) at the top where a cannon was mounted (usually a 32-pounder firing a 14.5-kg ball). Each Martello was a solid mass of 500,000 bricks, with chambers hollowed out for accommodation and stores. Its surface was a smooth coating of cement and sand stucco. The entrance was half-way up the landward side.

Martellos had interlocking fields of fire to prevent seaborne assault while the Royal Navy dealt with the warships offshore, and British troops rounded up any Frenchmen who survived – a classic example of the fleet-forts-field army strategy.

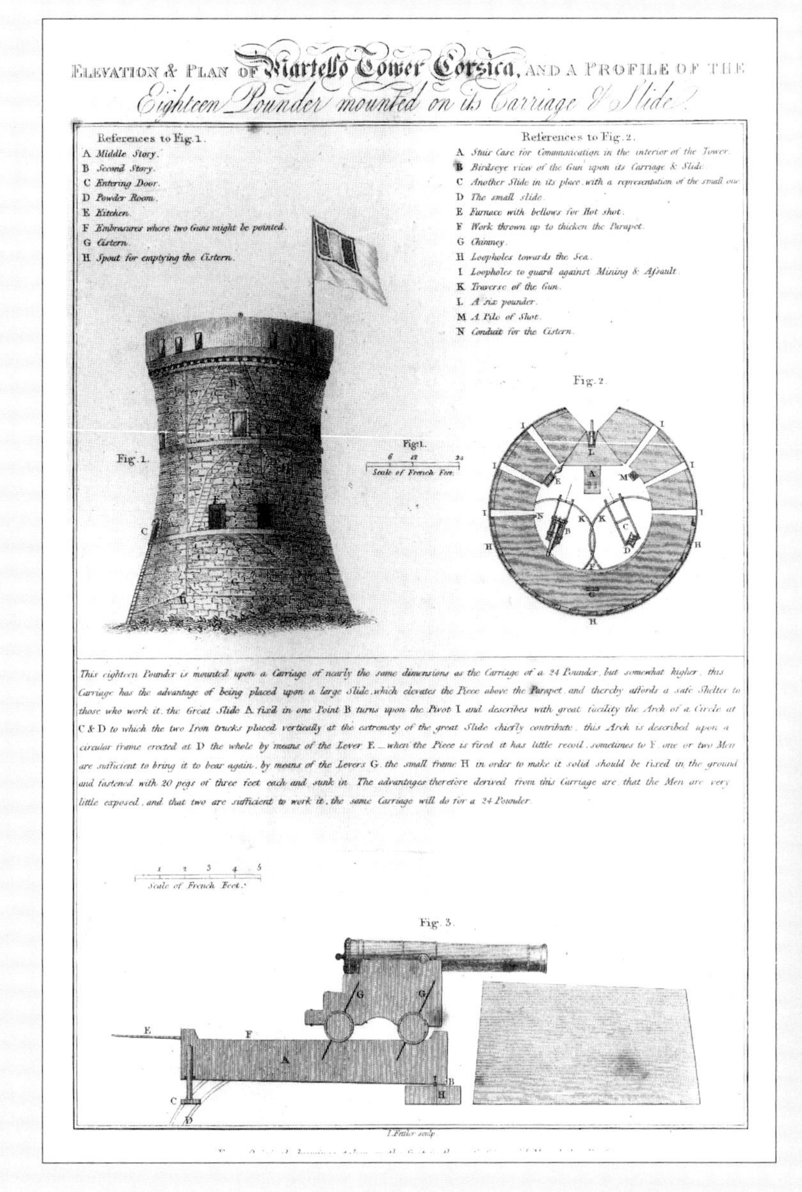

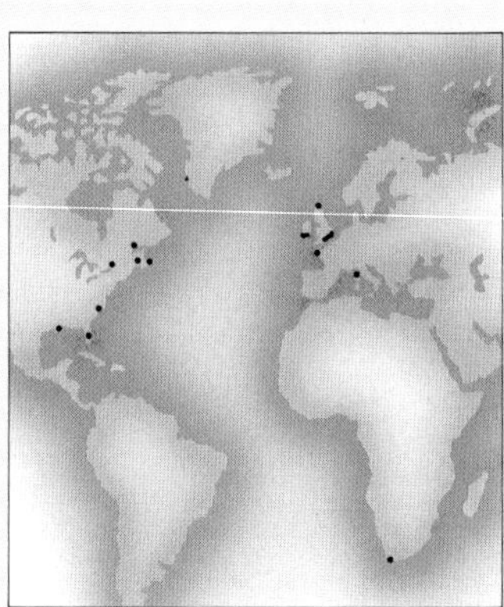

Left: A contemporary description in word and sketch of the original tower on Mortella Point, Corsica. The map **above** shows the areas where Martello Towers were built.

Below: Martello Tower No.64 at Pevensey Bay in Sussex.

Hostilities against Spanish forces did not end until 1659. Between that date and Vauban's death in 1707, there also occurred the War of Devolution over the Spanish Netherlands (1667-1668), the Dutch War (1672-1678), the War of the League of Augsburg (1688-1697), and the War of the Spanish Succession (1701-1713). During all these conflicts, Vauban brought discipline, organization and planning to the French army's siege operations, which were conducted under his direction.

At that time it was the common practice for besiegers to establish both a circumvallation (to prevent the garrison from breaking out) and a contravallation (for their own security). The besiegers then attacked a fortress obliquely, digging comparatively long, straight trenches, relying on their angle of approach to prevent enfilade fire. These trenches deepened as they neared the fortress, until they were perhaps as deep as the defensive ditch, and were themselves provided with firing steps and many other forms of fortification. They were also wide, to permit wagons and artillery to move up to the redoubts constructed as refuges in the event of a sortie by the garrison and as bases for the capture of land nearer the fortress for the further construction of more approach trenches and redoubts. Meanwhile the covering bombardment could be continuous, spectacular, and destructive – of the town within, if not always of its defences.

Most of these operations were conducted only during daylight hours, so it all took a long time, resulting in casualties from illness and combat, especially in the final assault which relied upon several columns of marching men pushing their way through whatever breaches had been made; the tail might still be back in camp while the head was entering the city. And at any moment the enemy's field army might arrive.

Vauban's new techniques were quicker and more efficient, enabling cities to be captured in days, rather than weeks or months. And not all fortresses had to be stormed. Once the wall had been breached, the city's capture was inevitable. So, having made a gallant stand and gained time for their own field army or the neighbouring fort to make preparations for battle, the defenders could surrender honourably. After all, during this age of Formalism, the aim of warfare was not now the extermination of the infidel or the heretic, but the calculated exercise of power over the chessboard of Europe. Fortresses were captured, only to be exchanged at the conference table for some far-distant colony, repossessed from a token garrison, dismantled, and then rebuilt.

Warfare of the late seventeenth and early eighteenth centuries might not have seemed much like a civilized game to the survivors of a storming party blown up by the explosion of a mine, nor to villagers watching their pillaged homes burn. But to their rulers it was as orderly as the formal gardens that surrounded their well-regulated palaces.

In addition to revitalizing siege warfare, Vauban was also a builder of fortifications. It is said that 150 fortified towns and lesser fortresses owe their defences to Vauban. That does not mean that he personally traced their lines and supervised the construction of each one, but it is certainly an indication of his authority – and of the wealth of resources available to the French army under Louis XIV.

Vauban's successors, even in France, could not be so liberal with money, men and materials. In any case, Vauban and his contemporaries had built so well, while muzzle-loading solid-shotted cannon had reached the limit of development, that there was hardly any need to design new types of military architecture. It was not until the beginning of the nineteenth century that there was any genuine development in military architecture in Europe.

But in the meantime, colonists in the savage wilderness had been compelled to construct strongholds – and to assault enemy fortresses – of a form quite different from anything Vauban ever knew.

Chapter Four

THE EXPANDING WORLD

The European wars of the sixteenth, seventeenth and eighteenth centuries were fought for reasons of princely prestige, religious freedom, national honour, community survival, and economic rivalry. None of these conflicts was fought *in vacuo*: neighbours and allies were dragged in or bribed to keep out until eventually campaigns were conducted wherever rival subjects encountered each other. Traders or missionaries, their interests were defended and advanced by armies and fleets until every part of the globe became a battleground. The result was that European influence and power extended over the whole Earth. There was no people, no territory, whose natural resources were not developed and exploited by European industrialists; and no society whose culture was unaffected by European civilization.

This was particularly so in military affairs. Foreign rulers of every kind, from Stone Age Red Indian Chiefs to the mighty Moguls of India, were quick to appreciate the advantages of European weaponry and organization – not that these other warriors were lacking in their own forms of armament, valour and discipline. And in the subject of military architecture, the European soldier found much to marvel at in the alien environments he was forced to journey through and fight in, from ruined citadels, their history forgotten and their ghosts feared by nearby villagers, to the teeming fortress-cities of India; from primitive stockades of murderous ingenuity, to the Great Wall of China.

EAST MEETS WEST

In general, the ancient civilizations of the Far East were able to come to some sort of understanding with the onset of European thought and technology – even before the introduction of firearms. It is recorded that in 1280, during Marco Polo's visit to China, he suggested the use of catapult artillery when Kublai Khan's army was besieging the city of Siang in Hupeh. The soldiers had no idea what the Italians meant, so the foreigners organized the construction of huge siege-engines that threw 140-kg (300-lb) stones right up and over the walls, hurtling down to smash roofs, spread hearth-fires and kill people. It was more than the citizens could bear, and they surrendered before their urban fortress was completely destroyed. But though effective, such a monstrous device did not really fit into the Chinese theory and practice of war, which was geared mainly to the employment of anti-personnel weapons and missiles; human beings were more expendable than cities! So when the Venetians travelled on, the Chinese chose to forget what had been demonstrated.

European firearms and cannon could not, however, be ignored by the Chinese from 1517 onwards. Even then, foreign artillery and fortification experts advising the Imperial Court found their proposals frustrated by the machinations of bureaucrats and eunuchs, jealous of a rival department's success in winning the Emperor's favour.

Sometimes it was necessary simply to accept that they had been outfaced by Western technology. That happened in 1860 during the 'Arrow' or Third China War. The Chinese government did not wait to see if their city walls and gates could withstand British and French bombardment, but surrendered just before the guns opened fire. Peking's walls might have stood up to considerable battering, although a general shelling of the city would have been a different matter.

The Manchus (who gained power in 1644) believed in keeping both foreigners and native-born Chinese separate from their sacred persons. Under their rule, the urban complex of Peking had been rebuilt in the form it retained until the twentieth century. Its entire defensive perimeter measured 27.4 km (17 miles), walled and moated. From the northern wall of the Chinese City extended the Tartar City, (or Inner City), its walls as much as 22.5 m (74 ft) thick and 12 m (40 ft) high. Within the Tartar City was the walled Imperial

Right: The pagoda-like form of the Castle of the White Heron. It stands on a hilltop at Himeji, between Kobe and Okayama on the Inland Sea coast of the Japanese island of Honshu.

City, the seat of governmental administration. And within that, walled and moated to prevent profane approach, was the Forbidden City, reserved only for the Manchu Imperial Family and their immediate entourage, and where no Chinese might enter.

During the nineteenth century, foreign legations were established in the Tartar City in a rectangular compound measuring some 800 × 400 m (880 × 440 yd). It backed against the wall dividing the Tartar and Chinese Cities. By manning both the parapet and flanking barricades cutting off this sector of this wall, plus the slighter wall surrounding the compound, the legation guards and desperate civilians held out for 55 days during the Boxer Rising of 1900. Luckily for them, neither the rioters nor the Imperial Palace soldiers were armed with anything heavier than swords, rifles and ceremonial cannon. But when the vanguard of the relief force fought its way through from Tientsin, they faced a predicament similar to that of the besiegers of the legations. Possessing only the small arms and ammunition they could carry, the advance party could not batter down the outer city gates and get through to the legations. So teams had to scale the uneven stonework or crawl through drains, then seize the fortress-gateways and admit their comrades waiting outside.

Among the soldiers, bluejackets and marines making up this international relief force were personnel from Japan.

The arrival in Japan of the first European visitors back in 1542 had coincided with a shift in political power there, as reflected in new forms of military architecture. Until then, fortification had comprised ditch and palisaded rampart, and various wooden buildings within. Each *daimyo* ('great name' or lord) had such a stronghold. There were about a thousand of them, professing allegiance to the Emperor but otherwise independent.

In theory the country was governed by the Imperial 'great general' or *shogun*. In practice this only happened when the *shogun* was powerful and ruthless – like Oda Nobunaga (1534-1582). He armed his followers with muskets introduced by the Europeans and overwhelmed the forts of lesser lords. He then began building stone castles, each garrisoned by his own men under a *daimyo* cowed into loyalty by fear, punitive tribute and hard physical labour in actual fortress-construction.

Shimonoseki Straits, 1864: new Japanese batteries bombard Western shipping. So on 5-6 September, they are captured and demolished.

SOUTHERN STRONGHOLDS

From the mystery of Great Zimbabwe to the magnificence of Machu Picchu the principles of good defensive architecture held good

Great Zimbabwe is about 250 km (150 miles) south of Harare. It seems that the ruins of a range of small houses in the valley are the remains of an ancient African city. On a 75-m (250-ft) granite knoll to the east is another complex of ruined walls, their cyclopean masonry fitted into naturally-located boulders and springing from the living rock.

The position of this complex suggests it was a fortress – although its exalted elevation, with a view towards the rising sun, could equally have had religious significance. Conversely, the situation of the elliptical building in the valley to the west of the ruined town implies that it has been regarded as a temple or palace rather than a fortification, although its layout and structure suggests a defensible citadel. The dry-stone wall of squared granite is 10.5-m (34 ft) high and 3-5 m (10-16 ft) thick, and it has a perimeter of almost 300 m (300 yd). Zigzag courses of stones near the dilapidated top part of this wall seem decorative rather than martial.

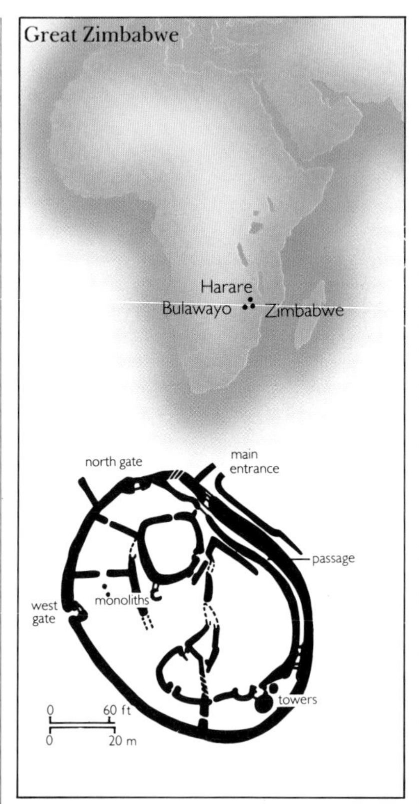

Above (top): Map showing the location of Zimbabwe.
Bottom: Plan of the Elliptical Building in Zimbabwe.

The main entrance is to the north, and has two approaches. One is short and direct; the other runs along a 30-m (30-yd) passage paralleling the main wall. If an assailant made it up the steps and through the gate, he then had a choice. He could turn right into a dead-end. He could go straight ahead, through a gateway that presented him with another triple choice of direction, all equally confusing and all leading him into complicated killing-grounds around a squarish fortification. Or he could turn left at the main gate and pass through a narrowing corridor until emerging into yet another multi-choice complex of fortifications. All this obfuscation was in addition to the other walls which divided the whole structure into precincts of varying shape and size.

Dates attributed to its foundation vary between the tenth and sixteenth centuries AD, although it may possibly date back to AD 600. Some enthusiasts claim that it is the site of the legendary King Solomon's Mines, or that it was erected by some long-lost Roman legion or by the Nubians, whose Sudan-based empire dominated ancient equatorial Africa. In fact, it was built by the ancestors of the Shona who live there now. Their country is named Zimbabwe after these timeless ruins.

Below: Though ruinous and overgrown, the walls of the Great Enclosure of ancient Zimbabwe still rise impressively from the soil of Africa.

Above: The Inca ruins of Machu Picchu in the Peruvian Andes.

Below: A VC was won attacking Gate Pah, on 28 April 1864.

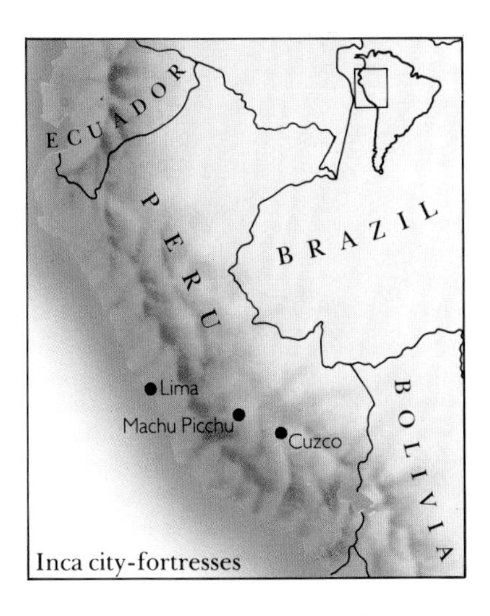

Above: Map showing the location of Inca fortresses mentioned in the text, in relation to modern boundaries. There were actually many more such strongholds than are shown here.

The Inca city-fortresses of Peru dominated the roads and suspension bridges climbing up from the Pacific coast into the Andean heartland. They also served as residential and temple complexes, relay stations for Imperial runners, tribute-collecting points, military bases, and refuges in time of trouble.

Machu Picchu, 100 km (80 miles) west of the Inca capital of Cuzco), was a city 2,500 m (8,000 ft) up in the Andes that covered an area of 11 hectares (27 acres). Its terraces were accessible only by staircases, each of which could be defended by just a few warriors.

Cuzco itself was overshadowed by the refuge-fortress of Sacsahuamán. Its blocks of stone are over 4 m (12 ft) in dimension, the walls rising 15 m (50 ft). Holding that, the Inca could have defied Pizarro for ever. But in 1533 they were preoccupied with civil war, and the Spanish *conquistadores* were able to capture Cuzco and seize its ruler Atahualpa for ransom before Sacsahuamán could be garrisoned.

Like Japanese and European medieval warfare, Maori campaigns were conducted according to a certain code of rules which was reflected in the construction of their strongholds, or *pas*, in the North Island of New Zealand, soon after their arrival in about AD 900. The earliest was at Te Awanga (Hawke's Bay). It was a timber palisade with interstices through which long wooden spears could be thrust to pierce attackers trying to force their way in. Then a terrace was built on the inside of the stockade so that the defenders could throw stones over the top. The earth on the outside of the palisade was scraped away, so that the stockade now stood on the edge of a steep little slope. A refinement of this ploy was to site the *pa* on the edge of a cliff. Then came the idea of rampart and ditch, similar to British hillforts. Multiple ditches and ramparts could be dug, and staging erected on top of the palisade from which missiles could be hurled upon the attackers.

Besides defensive structures, a *pa* contained stores of food in pits and huts; a well-stocked *pa* would be a tempting target for neighbouring clans after a season of bad harvests. A watchtower was manned night and day, and when the alarm was given, people in unfortified villages would hurry into the *pa* while the warriors prepared for battle.

When the war party arrived, it carried out a preliminary *haka*, or war dance, challenging the defenders to come out for single or group combat. The whole affair could thus be settled without an attack on the *pa* itself. If the challenge were not accepted a frontal attack was then carried out – a long siege was very unlikely. Sometimes the attack succeeded, sometimes it was driven off. Sometimes, however, the attackers pretended to run away, whereupon the defenders rushed out after them and a general mêlée resulted. Ownership of the *pa* was thus decided without its deliberate destruction. Indeed, its capture increased the victors' *mana*, or spiritual authority, and decreased the opponents' by a corresponding amount. An additional factor in this acquisition of moral strength, was ritual cannibalism after the battle.

Left: After World War II destruction, Kumamoto Castle (western Kyushu, Japan) was reconstructed exactly to its 1609 appearance. Curved walls withstand earthquake shock and the thrust of rain-sodden soil.

It was a policy continued by Nobunaga's successors, Toyotomi Hideyoshi (*kampaku* or civil dictator, 1582-1597) and Tokugawa Ieyasu (*shogun* 1597-1616).

Yedo (later Tokyo) was destined to be the capital of Japan, but Osaka (also in Honshu) was the greatest fortress of the period. First came natural water defences, then a ditch 11 m (36 ft) deep and 73 m (240 ft) wide. Beyond that rose a dry-stone wall of shaped cyclopean masonry. Then came another ditch and another wall, rising to 36 m (120 ft). Parapets, embrasures and the fortified gateway were armed with muskets, cannon and catapult artillery. The centrepiece was the *daimyo*'s tower-residence.

Defences were put to good use during the Christian rebellion against another Tokugawa *shogun*, Iyemitsu. Suffering religious persecution (partly prompted by Adams the Englishman), the Roman Catholic Japanese took shelter on the far west island of Amakusa. Hara Castle seems to have been one of the older neglected forts, but on 10 January, 1637 its ramparts were repaired and decorated with crosses and Christian banners. The 20,000 soldiers and 17,000 non-combatants were led by Masuda Shiro, a 17-year-old samurai who had been ordained a Christian priest (in the tradition of Buddhist warrior-priests). The besiegers totalled 106,000, including Dutch artillerymen and their warship.

The Christians repelled major assaults on 3 and 14 February, 1637 and were subjected to bombardment and steady attrition. However, on 11-12 April, 1637, one clan of besiegers tired of waiting and launched their own private night attack. The Christians, starving and out of ammunition, fell back. The assault gathered momentum until the whole army joined in. Traditions differ over the precise number of survivors, and even over whether there were any at all.

Certainly the Japanese felt that none of this would have happened if foreigners had not come to the islands. All aliens were therefore expelled – apart from a few Dutch traders as reward for their rather ineffectual and unwilling assistance at Amakusa. Japanese fortification now developed its own style, reflecting the same sort of society as had created the medieval castles of Europe.

There were two types. A *yamajiro* was a highland stronghold with a plain exterior befitting its harsher surroundings; it was of simpler construction than a *hirojiro* or lowland fortress. A typical *hirojiro* had a stone retaining wall around the base, with a concave face to resist the heavy thrust of rain-sodden soil and to distribute the stress of earthquake shock.

The pagoda-like towering levels of white walls and grey roofs, with considerable decoration, appeared massive structures, but in fact they were lightly built of resilient wood and plaster and merely sat upon the foundations instead of being fixed to them. They could thus be shaken by earth tremors and yet remain standing; more solid structures did not have such resilience and would collapse under similar shock. Within were stores of food, water and weapons, accommodation for the samurai knights, soldiers, servants and concubines, and private and official apartments.

The upper storeys of the *honmaru*, or keep, were fortified with arrow-slits, musket-loops and cannon-

DEBUL

One of the earliest occasions on which the use of catapult artillery was recorded in India occurred in AD 711 during the eastward expansion of the Muslim Arabs. While their main army marched overland through Baluchistan, the 'Bride' (with a crew of 500) was brought by ship to a rendezvous on the flat land of the Indus Delta. The objective was Debul, a fortified city defended by 4,000 warriors and 3,000 priests, who had erected a sacred flagstaff to make the city inviolable. The Arabs shortened the arm of their huge catapult to provide near-horizontal trajectory. The third rock brought down the talisman. The Rajputs should now have surrendered, but instead they launched an attack, were driven back, and were in turn assailed by escalade. Debul was taken, and all those who refused Islam were killed.

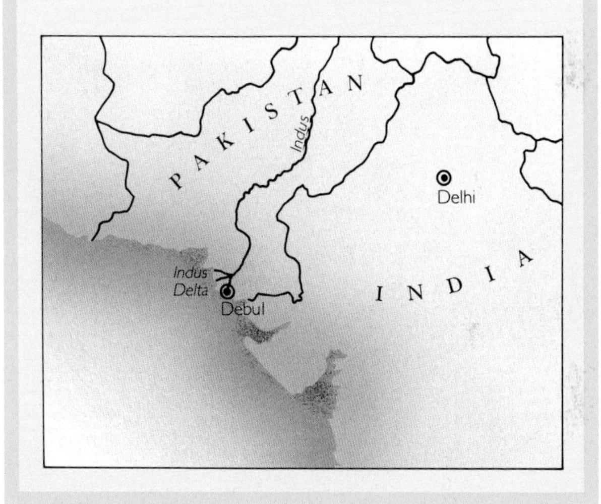

embrasures, as were the outer defences. These were laid out with drawbridges, bastions and dead-ends, affording good firing-positions and killing-grounds.

Nevertheless, Japanese warfare was traditionally settled in the field; samurai swords, barbed arrows and musket-balls killed the enemy but left his property and territory intact for the conqueror to occupy. As a result, sieges during this period of isolation seem to have been rare, and although very vulnerable to fire, most Japanese castles were spared military incendiarism. They were, however, sometimes afflicted by deliberate arson as part of ritual suicide following defeat.

It was only after the arrival of the American Commodore Perry, who forcibly reopened the islands to Western ideas, that Japanese castles began to suffer damage and destruction on a severe scale. This was inflicted partly by foreign warships as punishment of coastal batteries whose *daimyo* refused to recognize that a new era had dawned, and partly by Imperial troops. Awakened to the power of governance, the Emperor was now determined to rule the country *his* way. Woe betide any reactionary *daimyo* or samurai who tried to continue in the old ways of local, private warfare. Their defiant strongholds were smashed by the latest Western weaponry, served by soldiers trained in Western tactics.

Left: On 25 June 1859, the Taku forts at the mouth of the Peiho River prevented 22 British, French and American warships from advancing on Peking to force the Chinese Emperor to accept Western traders. On 21 August 1860, fourteen warships returned with 18,000 troops. After one Taku fort had been bombarded and captured, the others surrendered. The warships proceeded to land the rest of the soldiers, marines and bluejackets who marched on Peking.

Above: Another view of Kumamoto Castle in Japan. Shutters blank off ports for discharging stones, arrows and muskets, when not in use.

THE FALL OF MAGDALA

In a completely different part of the world, the Emperor of Abyssinia fared less fortunately in his dealings with Europe. Theodore III (proclaimed Emperor in 1853) opened correspondence with Queen Victoria, felt insulted by the British Foreign Office's delay in reply and by what was said when it did arrive, and subsequently imprisoned a number of Britons (and other nationals). These were released on the approach of General Napier's rescue expedition, but the assault on Theodore's stronghold of Magdala still went ahead.

Magdala is about 350 km (240 miles) from the coast, situated on top of a mountain, sheer cliffs on three sides. On the fourth, a narrow track led up to a massive gate in a 4-m (12-ft) wall. The armament included 15 cannon and a huge mortar.

At 9am on 13 April 1868, the attack went in. Covering fire was provided by two companies of the 33rd Regiment of Foot; the Royal Engineers and the Madras Sappers and Miners dashed forward to blow in the gate and lean scaling-ladders against the walls ready for another six companies of the 33rd to storm. But in the excitement, the explosives and ladders had been left back at the start-line. The engineers had brought some crowbars, but they made no impression. So, propelled from behind and beneath by rifle-butt, a drummer-boy was pushed up on to the top of the wall. He helped up a private soldier, who kept off the Abyssinians while the

AHMADNAGAR

CHAND BIBI was Queen of Bijapur in the late sixteenth century. Widowed and displaced during the expansionist policies of the Emperor Akbar, Chand Bibi took command of the garrison at Ahmadnagar, part of the Golconda-ruled Deccan about 240 km (150 miles) east of modern Bombay. Unlike many Indian fortresses, Ahmadnagar is situated on level ground, its walls of ashlar (or squarely-dressed stone) and brick. The main gate was defended by a semicircular barbican 10 m (30 ft) high, accessible only over a drawbridged floodable ditch, the whole approach necessitating four changes of direction.

The Moguls undertook their first incursion into Ahmadnagar territory under the General Mirza Khan after 1593. In due course, direct assault was made upon the walls, which were actually breached. The garrison abandoned hope and wanted to surrender but Chand Bibi put fresh heart into them; she personally led the counter-attack, and improvised repairs. The Imperial troops could not follow up their temporary success, and in 1596 Akbar Khan instructed his general to march away from Ahmadnagar.

In 1600 the Emperor himself arrived to take charge of a second siege. As it dragged on, Chand Bibi had fewer warriors to man the walls, and neither Bijapur nor Golconda sent a relieving force. On a rare day of deep depression, Chand Bibi for once expressed her fears aloud. Whereupon her already despondent bodyguard hacked her to pieces and handed over the fortress to the enemy.

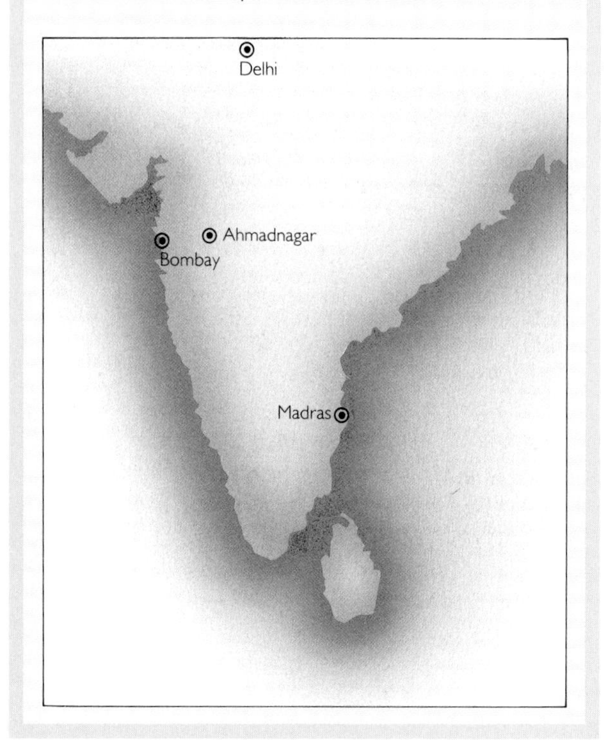

boy helped up an ensign (waving the regimental colours) and the rest of the 33rd. Emperor Theodore shot himself with Queen Victoria's presentation pistol, and Magdala was systematically demolished. 15 British soldiers had been wounded.

The fate of Magdala illustrates the fact that, unlike the ancient civilizations of the Far East, the less developed societies of Africa and America found it more difficult to withstand European militarism and culture.

Part-way between these types of civilization lay the Indian tribes of eastern North America, simultaneously employing both Stone Age technology and sophisticated systems of government. This is reflected in their military architecture which, because of their forest environment, was of wood.

THE NEW WORLD

Some early representations of such strongholds may have been artistically and imaginatively embellished or, conversely, stylized. Low slatted fences enclosing rectangular villages or gardens were undoubtedly just that: a simple barrier to prevent deer and other wildlife from trespassing on hutments and crops. Other barriers were decoys, funnelling game into killing-zones or

Magdala, 13 April 1868: even after General Napier's troops have toiled all the way up the mountain and past the circular gatehouse, they still have to rush Emperor Theodore's walled stronghold perched right on the edge of the cliff.

trapping enclosures. Circular fences, however, suggest more warlike purposes, especially where one 'wall' overlaps the other to form an extended, constricted gateway approach. That was the case at the Indian town of Pomeiooc in Virginia. Its vertical posts were about 2-3 m (6-10 ft) high and slightly spaced so that defenders could stab through at attackers trying to scale the obstacle.

The fortified village of Hochelaga (where Montreal now stands) was a much more complex affair. It was, after all, one of the principal cities of the Iroquois Confederation. Planks were fixed together to give the circular wooden wall a cross-section in the form of

HIGHLAND FORTS

It is not often appreciated now that until the beginning of the nineteenth century the Scottish Highlands were regarded as as wild a frontier as any Red Indian territory. To maintain the peace in this raw region, forts were built at strategic points. One such was established at the southern end of the Great Glen by General Monk in 1654-5. It was rebuilt under William III and called Maryborough until taking its present name of Fort William. It withstood Jacobite siege in 1715, but underwent further strengthening according to the current ideas in fortification, as part of General George Wade's programme. Between 1726 and 1740, this English army officer organized the construction of a network of military roads linking strategically-placed forts.

The Great Glen was blocked by Fort William itself, by Fort Augustus in the middle (improved in 1730), and by Fort George (where Inverness Castle now stands). Ruthven Barracks (near Kingussie; built 1716, enlarged 1734) and Bernera Barracks (opposite Skye; built 1722) are reminders that these places were depot-bases for police-patrol rather than castles. In fact, when the 1745 Rebellion came, most were taken and destroyed by the Jacobites, although Bernera and Fort William held out (perhaps because of their maritime situation). Wade's roads came into their own in the aftermath of Culloden, and so did the forts refurbished by Wade's successor, General Caulfield.

Highland Forts and Military Roads

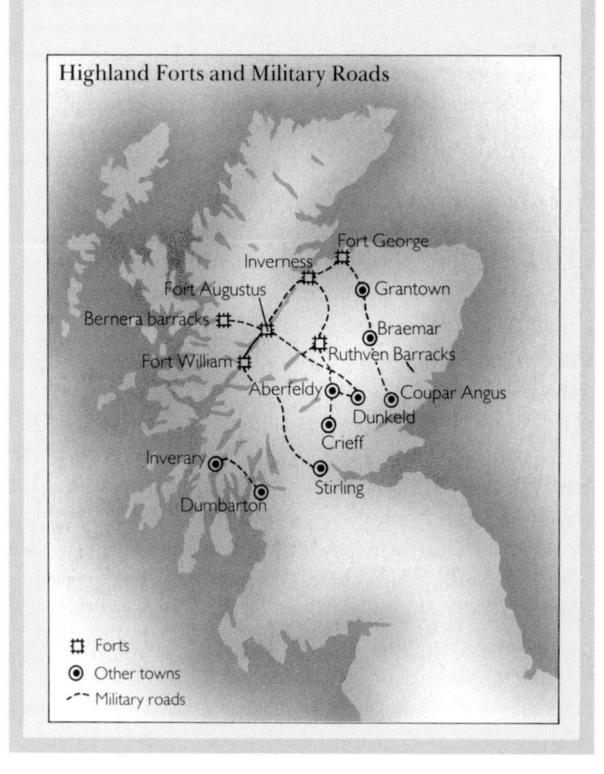

AMERICAN FRONTIER FORTS

From the stockades of invading colonial powers to the outposts of the Spread to the West, the American strongholds guarded the new frontiers

In the beginning the shore-line was the frontier. A reconstruction of James Fort, established in Virginia in 1607 by Christopher Newport and Captain John Smith, shows a triangular fortification. The stockade facing the river was 128 m (140 yd) long, the other two sides 91 m (100 yd), their apex facing inland. Each corner had an earth-filled, wood-revetted round bastion, enabling the cannon mounted there to cover the front of the walls. A cannon also stood in the centre of the fort, its muzzle threatening the gate in the river-wall in case of hostile entrance. There was a church, a well, and small houses reflecting the simpler domestic architecture of the England the 103 settlers had left behind. The use of earth and timber in this fortification reflects the principal materials available on this alluvial plain.

The Spanish, farther south in Florida, were accustomed to working in stone for their homeland castles and for their fortifications in Central and South America. When suitable local stone was not available, they either used adobe (blocks of sun-dried clay) or shipped good-quality building stone from the Iberian Peninsula. One exception was San Marcos Fort (by Saint Augustine in Florida), built of workable yet durable rock from the seabed, into which cannon-shot thudded harmlessly without shattering.

However, it was in the higher latitudes of North America, with their forests of tall, straight conifers, that timber fortifications were most developed. A 1546 map of the St Lawrence includes a sketch of an early fort established by Jacques Cartier. Evidently perched on the edge of the inaccessible bluff where Quebec now stands (and hence secure from attack on that side), it is a semicircular stockade of vertical posts. There is a simple but stout door, and embrasures for four field-pieces.

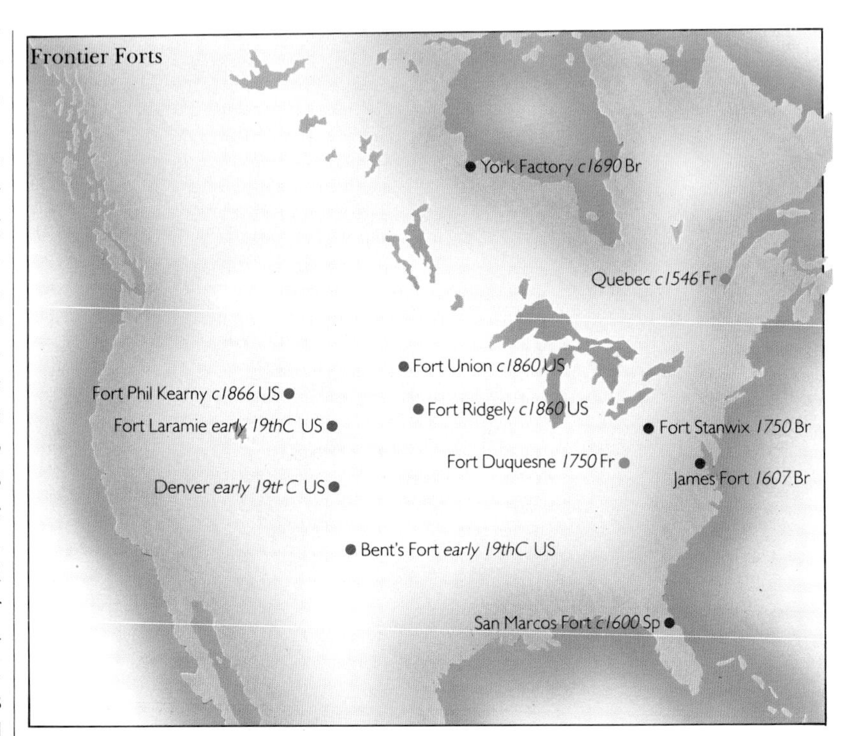

This map shows how Europeans first established fortified settlements on the east coast of North America (including Hudson Bay). Duquesne and Stanwix illustrate rival attempts to control the St Lawrence-Ohio-Mississippi river systems, thus inhibiting hostile forces or exploiting their own westward expansion. This last is represented by the private-enterprise fortified trading posts and government-manned military bases of the nineteenth-century USA.

By 1613, Quebec is shown as a collection of three large vertically-planked buildings comprising living accommodation, armoury and workshops. These are linked by an encircling upper-storey verandah, its solid parapet pierced with loopholes. This is continued over the plank walls of a triangular open area in which is a dovecot (for fresh meat). There is also a gateway with a *pont-levis* (drawbridge) across a ditch. Beyond that are triangular bastions of revetted earth for cannon. And beyond that again, is the garden for herbs and fresh vegetables.

By 1690, when Sir William Phipps led the Boston Militia in an unsuccessful amphibious assault on Quebec, the place had become the administrative capital of the French province of Canada. The original cliff-edge stockade had become the walled Upper Town, its public buildings approached via imposing gateways. The fortifications needed frequent repair, but the place was hardly a frontier fort any more. The frontier had moved on to places like York Factory, where the Nelson River flows into Hudson's Bay.

York Factory was named in honour of the Duke of York (one of the patrons of the Hudson's Bay Company) and because it was managed by a 'factor' – the company's local representative. Fort Nelson (which was its alternative name) was a wooden stockade, the tops of its vertical tree-trunks so trimmed that they formed V-shaped merlons and crenels. The walls also had embrasures. There were three lines of permanent stockaded breastworks between the water's edge and the fort, and any outflanking movement through the woods around the foot of the bluff was hindered by a shallow ditch and low earthen parapet. But although such fortification was effective in withstanding Indian or Eskimo attack, a surprise assault by a concentrated force of disciplined regular troops was a different matter. The small Hudson's Bay Company garrisons were traders first and soldiers second; York Factory was one of those outposts that fell to the French in 1697.

The most efficient frontier forts of this period were those designed on contemporary European lines, even if the building materials were of local origin and did not include stone. Two examples of this policy are Forts Duquesne and Stanwix, built by the French and the Anglo-Americans in the 1750s. Both

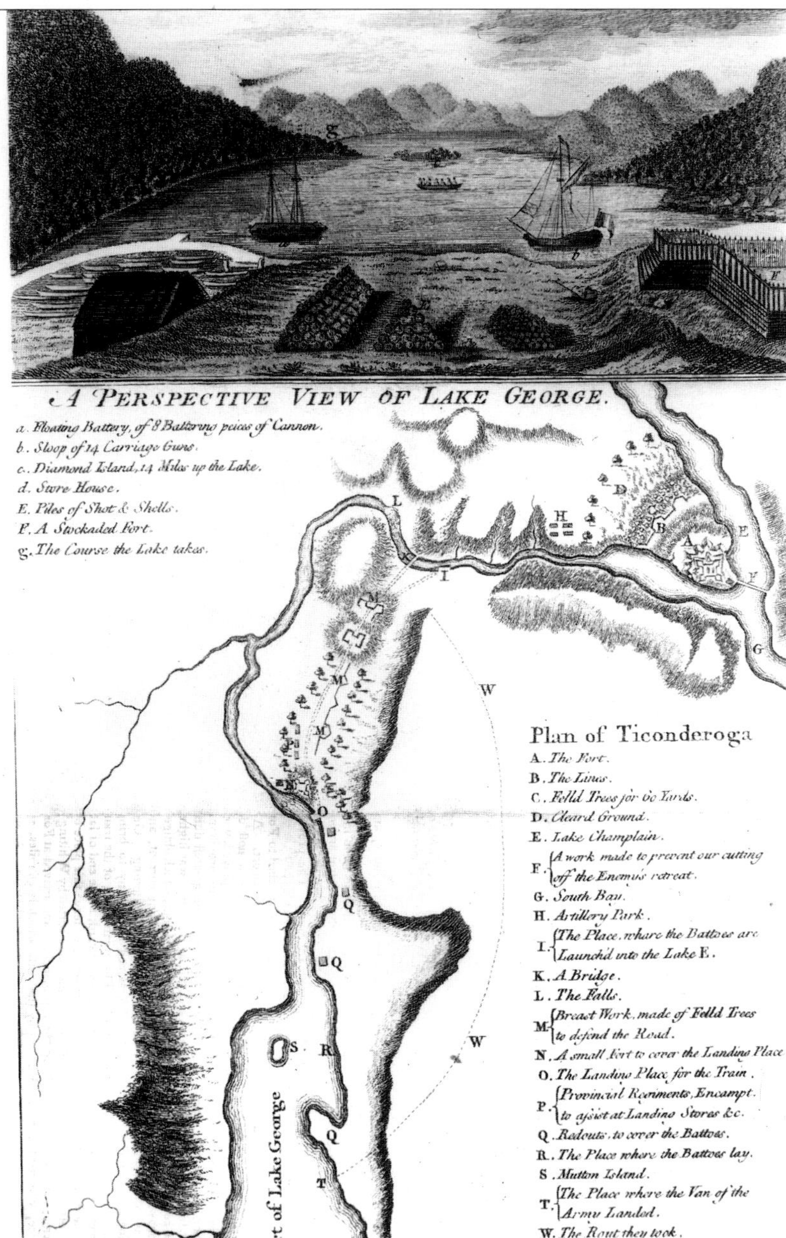

Above left: Splayed embrasures in the Citadel at Quebec enable cannon to be traversed to cover the St Lawrence. Musket loops and occasional gunports for grapeshot, in both Citadel and counterscarp walls, would eliminate any infantry who got into the ditch.

Left: 1777: Ticonderoga (orginally French, now garrisoned by Americans) is captured by the British.

dominated crucial river communications, and both acted as bases for operations in their respective areas. For example, Captain Dumas led 900 French and Indian troops out from Fort Duquesne to ambush and defeat General Braddock's 1,400-strong expedition across the Monogahela in 1755. However, William Pitt's distant strategy later began to take effect, and in 1758 Fort Duquesne fell. It was renamed Fort Pitt (and later became Pittsburgh).

Both Fort Duquesne and Fort Stanwix were square in plan, their four corners extended in lozenge-shaped bastions. A detail of a drawing of Stanwix shows that the walls between the bastions were made of horizontal timbers facing earth banks. Gun-ports and musket-loops were cut at ground level, with other cannon protruding over the bastion

Above: Fort Union on the Missouri River in Montana, pictured in 1833.

Below: Arx Carolina, a seventeenth-century Spanish settlement in Florida.

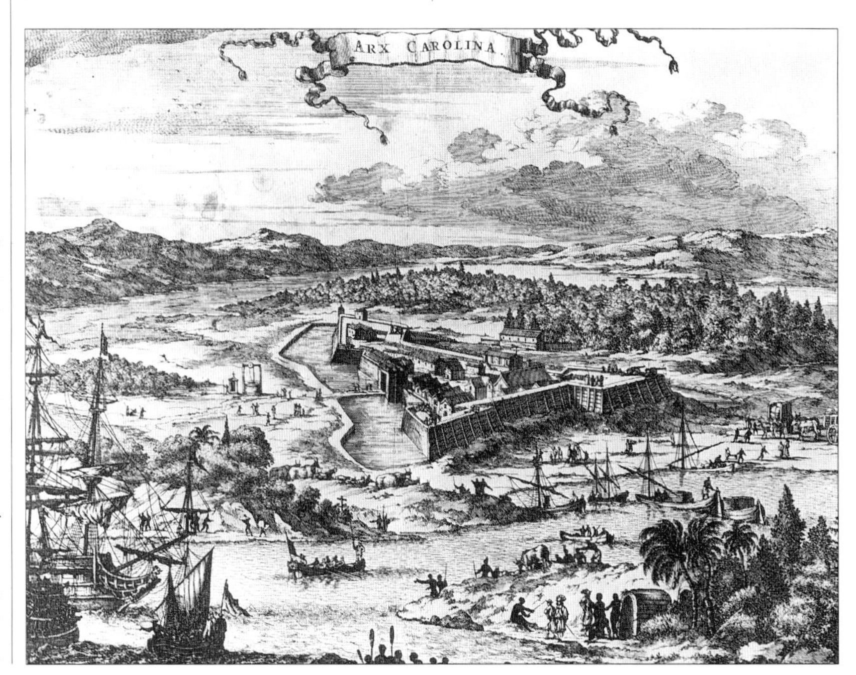

parapets. A square blockhouse guarded the bridge over the ditch.

These blockhouses were a feature of American frontier fortification; sometimes they were built alone and sometimes in association with other works. They were of horizontal timbers, thick enough to keep out musket balls. Rectangular or square, their ground-floor plan dimensions were about 6-10 m (20-30 ft). The groundfloor had four cannon-ports and a row of musket-loopholes. A ladder led up through a trap-door to the upper storey, which was slightly larger in area so that it marginally overhung the bottom level. The officers' quarters were up here, and there were more loopholes and gun-ports, but for lighter pieces.

There were also holes in the floor to fire down at anybody who managed to break in and occupy that level. Above this floor, and making a total height of 10 m (35 ft), was the roof – steeply pitched for snow clearance and to deny lodgement to flaming arrows and fire-brands. (The defenders must have hoped that no incendiary would lodge behind the chimney, an essential feature because a fireplace was indispensable for food and warmth during the winter.)

These blockhouses could accommodate 100-120 men in an emergency, although usually the garrison would be fewer, and even then some of the men would be out on patrol, sleeping rough under the sky. Meanwhile the blockhouse – if it had been well sited – could through its cannon deny riverine or road passage to hostile traffic.

The nineteenth-century cavalry forts of the far west likewise acted as bases for patrols in hostile territory. In addition they provided escorts for transcontinental wagon trains. They also served as refuges for those same civilian pioneers, and inevitably became centres of government administration. A typical cavalry fort measured 183 × 214 m (200 × 266 yd). If each suitable tree is assumed to be about 30 cms (1 ft) in diameter, the walls alone required almost 3,000 3-m (10-ft) lengths of vertical timber, quite apart from the timber for all the stables, accommodation, offices, stores, magazine, bandstand and flagpole. There were also the corner blockhouses – at least one, sometimes open at the top to act as command post. The cannon were lined up facing the gate: it looked orderly; it impressed 'friendly' Indians; and their fire would devastate 'hostiles' if they did manage to break in.

Below (top): A timber-built blockhouse at Fort Abraham Lincoln, the North Dakota base for Colonel Custer's ill-fated campaign which ended at Little Bighorn on 25 June 1876.

Below (bottom): US officer's sketch of Bent's Old Fort in Colorado. Built in 1833-4 as a fur-trading post, it eventually became a military base.

It was blasts from such obsolete instructional cannon that saved Fort Ridgely, in Minnesota, on 20-21 August, 1862, although it is true that that place had been a camp with dispersed buildings, a fort in name only.

It was less likely that Indians would be able to storm Fort Phil Kearny in Wyoming. Construction began on 15 July, 1866. It had a good supply of water and forage, and two neighbouring hillocks for lookout posts. There was plenty of timber, but not immediately around the fort. That was a good point because it meant there was no cover for Indians; it was also a bad point because soldiers had to be sent out to cut wood – and might well be cut off themselves, as happened on a number of occasions. But eventually Fort Phil Kearny was completed on 31 October 1866 – and was promptly invested by the Sioux under Red Cloud's leadership.

They did not attempt a costly storming of the fort, nor even the steady attrition of close siege, but their winter blockade took effect: the battles and privations became legendary on both sides. And even in the summer months, the garrison had to expend all its energy in maintaining its own operational viability, with nothing to spare for protecting wagon trains. Accordingly the Bozeman Trail was shut down. And on 2 August 1868, the US Army marched out, Red Cloud led his warriors in, and Fort Phil Kearny was reduced to ashes blowing across the land. It was the only time the Plains Indians engaged in a protracted, coordinated investment, and it was their most successful campaign. It turned out to be their last strategic victory.

Many of the remaining frontier forts became redundant when the Indian Wars ended, but others became centres for military training. During the later twentieth century, the remote security, level plains and dry atmosphere of the desert areas made them ideal locations for testing nuclear weapons, aircraft, rockets and spacecraft. No matter how sophisticated their equipment, the military establishments there are but collections of accommodation and workshops, encircled by thin barbed wire: they are the direct descendants of the frontier forts of America, fortress-bases for the exploration and exploitation of the New Frontier of Space.

A
B
C
T.B
19

the capital letter A about 10 m (12 ft) high. A shelf or wallwalk ran all the way round the inside, braced underneath and reached by ladder.

During campaigns against the French and their allies the Huron and Algonquin, the Iroquois improvised similar near-circular structures. Panels of horizontal logs lashed to lopped branches were fixed in the ground making a multifaceted stockade, from which they emerged to raid all the homesteads in the area, and into which they retired when outnumbered. As at the Battle of Hastings in England in 1066, attackers aimed their bows skywards, giving their arrows deadly plunging fire. Meanwhile, other besiegers were dislodging the logs under cover of birch-bark or hide mantlets. Another means of entry was to cut down a large tree so that its fall smashed down a section of the stockade. However, the Iroquois were not only savage fighters, they were also organized and disciplined, used to waiting for a well-timed sortie to bring victory.

It is often forgotten that Tsarist Russia had a frontier every bit as wild as the American West. To protect lines of communication along the Ural and Tobol Rivers, and thence to Kuznetsk in Siberia (reached by the seventeenth century), ditches were dug backed by a palisade of pointed stakes. The *Cherta* linked stockaded towns and natural barriers, and covered a total distance of 3,000 km (2,000 miles). Other similarly protected towns held garrisons in reserve who could act as a mobile defence in depth. The soldiers guarding the *Cherta* itself were accommodated in square, pointed-roofed timber blockhouses, which acted as watchtowers and forts, warding off Tartar horsemen.

The Mongols could of course have resorted to incendiary tactics, which would have been particularly dangerous in winter when water freezes before it can reach the flames. Nevertheless, fire is not a practicable weapon in such circumstances: it destroys buildings – potential sources of shelter, warmth, food and military material – as well as killing the people inside. Conversely, if the defenders could keep the property reasonably weatherproof and hold out long enough in comparative warmth, the attackers outside would die of exposure anyway.

In more primitive times, the only way of gaining admission to a timber stronghold might be to pose as a friendly traveller in need of assistance. Once inside he would assassinate the chief and open the door to his friends outside – a powerful reason for regarding all strangers with suspicion, a trait that could become deeply ingrained in a people's attitudes.

TROPICAL FORTIFICATIONS

Such considerations naturally do not apply in tropical regions in which timber grows luxuriantly, and where fire might only be employed on a large and terrifying scale in the dry season. In both 1824 and 1852, the approaches to Rangoon near the mouth of the Irrawaddy in Burma were protected by stockaded forts. Their enceintes traced right-angled salients and re-entrants, and the parapet and wallwalks sprouted banner-waving flagpoles. However, they afforded little protection against British naval guns firing at the maximum elevation of the day over the top of the palisades to penetrate the magazines. The Burmese found the stone complexes around temple-pagodas more effective for their last-ditch stands against the British invaders.

But there were further obstacles around fortified villages in Burma, and especially around the remote stronghold of the bandit-patriot Myat Toon. These included holes dug and camouflaged. The future

Left: Scenes of life in and around a Red Indian stockaded village in early Virginia. Though stylized, the depiction of agriculture and the details of domestic and community architecture suggest an advanced social structure.
Right: Old Cemetery Gate, Fort Canning, Forbidden Hill, Singapore. In the nineteenth century Raffles chose this area as Government Hill, to allay fears that it had been haunted since the 1391 end of the ancient Singapura stronghold.

FORT ST GEORGE

A Dual-purpose coastal-defence and inland-facing fort at Madras, on the eastern shore of India, representative of the strongholds built by European powers throughout the world, Fort St George was originally two concentric rectangular fortifications, the inner with arrowhead corner bastions. First built by the British in 1639, it survived a three-month siege by the Mogul General Daud Khan in 1702, and an attempted assault by a Mahratta army in 1741.

In 1746 its 200-man garrison was besieged by 3,700 troops and nine ships under Comte de la Bourdonnais. The bombardment began on 14 September 1746 and lasted until the fort's surrender on 25 September. Two years later it was exchanged for Louisburg, captured by the British on Cape Breton Island off Canada – a typical example of the treaty-deals of the Age of Formalism.

The French under Comte de Lally-Tollendaal were back in December 1758, but the garrison held out until February 1759, when a British fleet hove in sight and the French withdrew.

Fort St George was then rebuilt on a pentagonal trace with pointed bastions, an elongated hornwork screening the Sea-Gate, an extra internal wall along the north (the direction in which lay the greatest threat), a ditch all the way round (part flooded by the river), seven ravelins, and further earth outworks forming a 12-pointed enceinte.

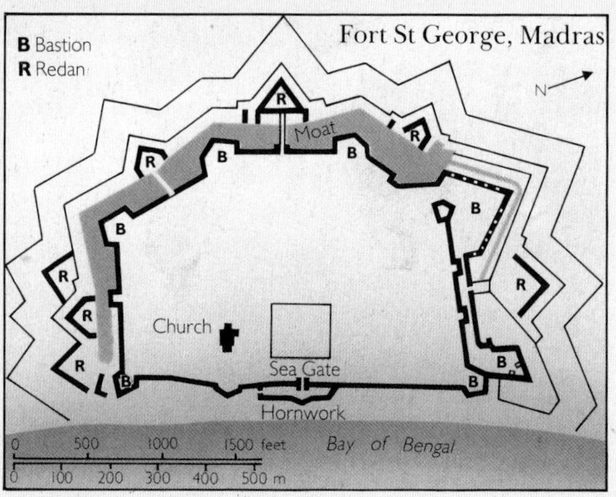

Above: The earlier Fort St George at Madras. The inner wall surrounding the Governor's House can just be seen, plus the church, offices, arsenal and barracks between the concentic walls. According to some accounts, the outer wall was provided with a round bastion at each corner. Here, the only visible corner bastion is square, with an older machicolated round tower above it. However, there is another multi-sided bastion on this eastern wall, plus a semicircular barbican fronting the Sea Gate. Note the emphasis on the flag, symbol of sovereignty in a foreign land.

Left: Plan of Fort St George in eastern India, also called the Coromandel Coast or the Carnatic. (Not far from Madras was the later base of Bangalore, giving its name to the twentieth-century device for blowing up barbed wire.)

Field-Marshal Sir Garnet Wolseley knocked himself out by falling into one in March, 1853.

Many primitive centres of governance or outlawry had titles more grandiose than was justified by their actual defences. This was so in the case of the 'Palace of the Buddhist King of Burma' in 1930. Saya San launched his rebellion against the British Empire on Christmas Eve. His army murdered a forestry officer near Tharrawaddy and burned any isolated Burman community that refused to pay tribute. Saya San confidently advertised his hilltop fortress by illuminating it with lanterns at night. But when a Kachin detachment of the Burma Rifles stormed it, they found the 'palace-stronghold' was a single leaf-thatched hut, and the 'army' totalled 200. Most of them, including Saya San did not stay to see whether their 'magic pills of invulnerability' worked; those who did were killed.

Such a campaign that turned out to be against a non-existent fortification is a reminder that European forces operating in alien environments could never take risks. Precautions had to be taken against possible enemy defence-works and possible hostile attack.

Before the Burma Rifles moved against Saya San's 'stronghold' they first established a tented base camp at Hlenglu. Its 300-m (310-yd) circular perimeter was marked by a palisade of sharpened bamboo stakes leaning outwards. Lewis-gun positions were continuously manned, and the barbed-wire gate was opened only when authorized personnel were entering or leaving, and then immediately closed again. Furthermore, only a couple of dozen riflemen went out on patrol at any one time; the rest of the unit stayed behind, some guaranteeing base security, while others stood ready to depart in support of their comrades.

Different environments demanded different measures. For example, the British army established overnight perimeter camps when on the march on the North-West Frontier of India. These were about 30 m (100 yd) square, and were composed of a shallow ditch and stone wall. The infantry battalion bivouacked just inside the perimeter; the transport and everything and everyone else then organized themselves within that screen of armed men, who took shifts to stay on watch. The greatest danger was likely to come from a sudden and sweeping attack in an attempt to storm the

Below: Former symbol of far-flung imperial pride and authority, a coat of arms embellishes Isabel Gate into Fort Santiago, part of the old Spanish defences of Manila in the Philippines.

GIBRALTAR

Through the centuries Gibraltar has proved a sure guardian of the gateway to the Mediterranean

Gibraltar illustrates how one fortress can expand its role for service in a changing world.

The Ancient World

The two Pillars of Hercules (of which the Rock is one) are held to mark the western limit of the Mediterranean world. The Romans called the Rock, Mons Calpe.

AD 27 April, 711

Tariq's North African Moors land on the isthmus joining the Rock to the mainland of Europe. They march into Spain. Their galleys, operating from the sheltered beach, strike at Visigothic targets all round the Iberian Peninsula. To protect his base from retaliation and to prevent Christian occupation and interference with the cross-Straits build-up of Muslim forces, Tariq builds a

Gibraltar's topographical and political situation inhibits military – and civil – expansion. Any development has to be by sea reclamation, underground tunnels or technical improvement of existing positions. The most significant differences since the drawing of the plan **below left**, have been enhanced dockyard facilities and a single-runway airfield on the low-lying isthmus, seen **above**.

Below: The addition of a handwheel and vertical screw instead of a wedge made for easier and more accurate elevation and depression of eighteenth-century cannon.

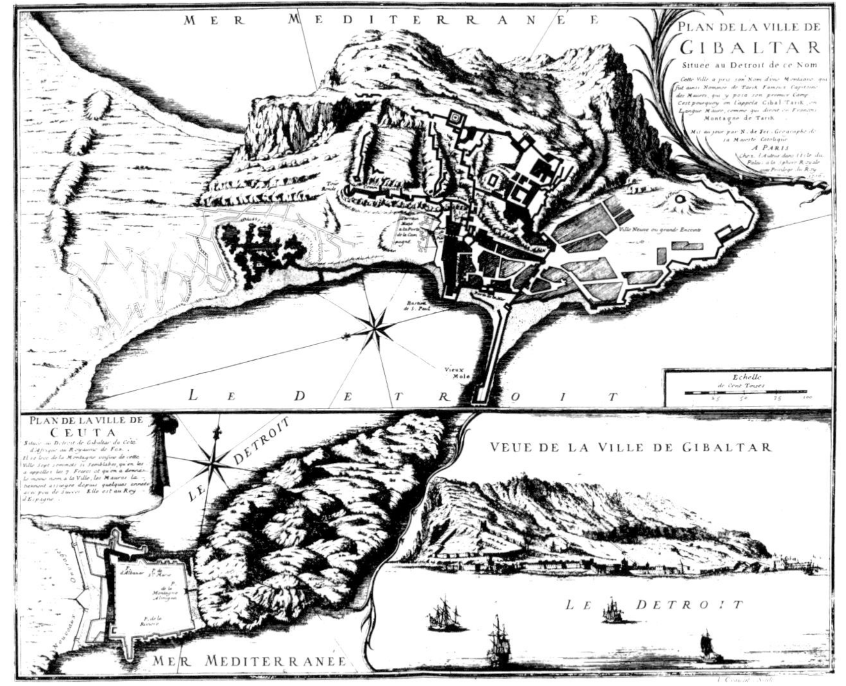

castle on the Rock overlooking his original landing beach; the Rock becomes known as Jebel Tariq.
962
The Spanish revival begins, but far away from Jebel Tariq.
1309
A Spanish army at last operates far enough south to capture Jebel Tariq, but the Moors still occupy much of Granada.
1333
Moors recapture Gibraltar.
29 August, 1462
Spanish capture Jebel Tariq. 30 years later Columbus discovers America. The Atlantic becomes the great highway to both the New World and the Orient. Gibraltar is vital for the protection of ships passing between the Mediterranean and the Atlantic. The Italian architect Giambattista Antonelli (1531-1588) repairs old and builds new fortifications.
1704
By now there are five sets of walls. They surround the Old Town to the west of the Rock; they run in parallel from there out along the Old Mole; they form the rectangular perimeter of a fortified area farther up the hill, which includes square citadels and round bastions; they extend north towards the flat isthmus in a series of right-angled and arrowhead salients and re-entrants, ending in a round tower; and, longest in extent, the Grande Enceinte follows the shore around to Europa Point and back. However, the Spanish government has neglected the defences; the 150-man garrison has only a few guns from a small French privateer.
21-24 July, 1704
Vice-Admiral Sir George Rooke lands 1,800 British and Dutch soldiers and marines to occupy the isthmus, while at least 20 warships bombard the fortifications until the Spanish surrender. From now on, Gibraltar is British, but that is not a foregone conclusion. There is no safe anchorage or dockyard for large sailing ships, nor enough water for the fleet. Until these deficiencies are rectified, the Royal Navy sails home every winter, leaving only a token presence in the Mediterranean.
October, 1704-April, 1705
The British in Gibraltar are blockaded by French ships and besieged by Spanish and French troops. 18 Royal Marines defend the Round Tower so valiantly that 'Gibraltar' becomes their sole battle honour.
1725-1727
The Spanish dig entrenchments across the isthmus, but the siege is more of a formality than a battle; diplomatic negotiations continue at various distant courts.
24 June, 1779-12 March, 1783
French and Spanish forces (including specially protected floating batteries) besiege Gibraltar by sea and by land. They are unable to maintain close naval blockade, so the Royal Navy is able to slip in from time to time with reinforcements and supplies.
Nineteenth century
Thanks to the Suez Canal, the Mediterranean becomes the great highway to the Orient, while the Atlantic remains vital for British communications throughout the rest of the Empire. Dockyard and coaling facilities are built for steam ironclads, while defences and armament are updated, including casemates and a 100-ton rifled muzzle-loading gun with a range of 12 km (8 miles).
World War II
Air base and radio station built to reinforce naval protection of Atlantic and Mediterranean convoys, and as springboard for invasion of North Africa. Armament now almost 100 guns, from 40-mm AA Bofors to 233 mm (9.2 in) Mark X with a range of 25 km (15 miles). More than 100 tunnels are excavated for accommodation, hospitals, stores, and secret sally-ports. A German assault on Gibraltar is abandoned because of the difficulties of operating in Spain. There are some air raids, but the main danger comes from Italian frogmen based in Spanish territory.
Post-1945
Some updating of armament, but guns are becoming redundant and are removed; development of radar and radio as part of worldwide surveillance and communications networks.
The Future
Will Gibraltar remain British or will it revert to Spain? Like so many other fortresses throughout history, Gibraltar is still a symbol of national sovereignty.

Montagu(e) Bastion, built on the north-east corner of the Rock for muzzle-loading cannon: during World War II, it mounted two 3.7-inch (94-mm) anti-aircraft guns. Underneath was an oil storage depot.

JALALABAD

A LONE horseman, staggering out of the winter mountains, gasped out the news that the whole of the British army in Afghanistan had been wiped out; he was the only survivor.

That was the dramatic prelude to the siege of Jalalabad on 13 January, 1842. General Sir Robert Sale's force officially comprised the 13th (Somerset) Light Infantry (700 men) plus 1,300 from the 35th Native Infantry and some Royal Engineers. But most were already ill or injured, and if the whole army could not manage evacuation, what chance did they have? So they decided to stay, repair the ruinous round-bastioned walls and citadel, and emplace their few cannon. This they did, only for the work to be undone by earthquake. Again they laboured.

The Afghan army arrived on 11 March, 1842. The first onslaught was beaten off, so Muhammad Akbar Khan organized a series of limited assaults to wear down the garrison. Not only were these repulsed, but the defenders made several sorties including resolving the food shortage by capturing 500 sheep on 1 April, 1842. The climax came on 7 April, 1842, when the garrison launched a counter-offensive and captured the Afghans' siegeworks and their camp, containing all their stores and equipment. The enemy was forced to abandon the siege and begin a retreat back to Kabul.

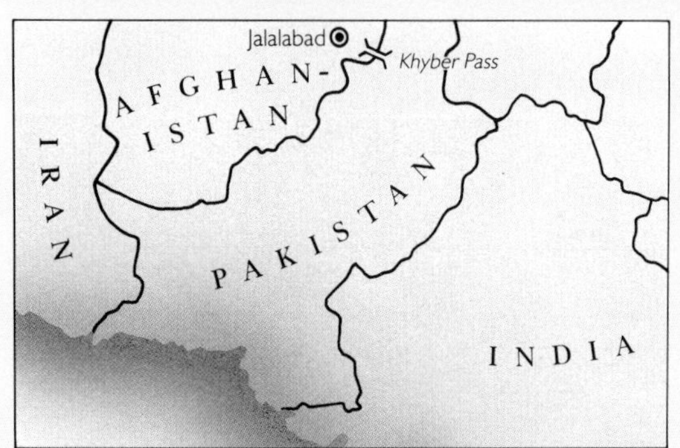

Above: During Queen Victoria's reign, the British Army in India fought some fifty campaigns on the North-West Frontier, conflicts which continued on through the twentieth century. Casualties varied from single sentries picked off by hidden snipers high up on the hillside, to the ambush and destruction of whole armies. In the 1920s-30s, the RAF attempted to solve the problem by bombing village-strongholds (as above) – but they could do little about the snipers. The Russians faced similar problems during their 1980s campaign.

Left: Map showing Jalalabad's location on the Afghan side of the Khyber Pass, the only really practicable route between India and Afghanistan – and on into Russia.

wall, and to pass right through the camp and out the other side. When an alarm was sounded, the infantry could man the wall without having to disentangle themselves from the rest of the camp. To cut down on sniping, an outer ring of camp picquets (each a platoon strong) was established at places that were potential sniper positions. Given provisions, ammunition and bedding, the picquets were protected by a stone breastwork (or *sangar*), plus a light trip-wire.

This type of fortification was essentially a fieldwork, but it could be developed into a permanent picquet. **During the Razani Campaign of 1937, the 1st/2nd** Gurkhas protected a road with four strong, heavily-wired picquets, each holding a striking force of one platoon which patrolled the surrounding slopes by day and by night, ambushing snipers or trailing them back to their villages and shooting them up. If the Gurkhas got into difficulties they could fall back to the picquet, which was continually garrisoned by a machine-gun section and – sometimes – a mountain artillery-piece.

When European adventurers began their exploration and attempted conquest of the rest of the world, it was inevitable that the permanent fortresses they built should reflect those same functions that characterized the castles in their own homelands. They served as bases for the military, administrative, commercial and cultural exploitation of the surrounding area; they were intended to impress and deter the local inhabitants; and they acted as refuges in time of trouble. The actual construction of these strongholds was subject to three main influences; the local style of building, arrived at for climatological or other reasons; the availability – or otherwise – of materials, labour and time; and the current trend in military architecture back in Europe.

IN DESERT LANDS

All these factors played their part in the nineteenth-century construction of small forts for the French Foreign Legion in the North African desert. Accommodating a garrison of 20-25 men, they were usually located near an oasis, itself invariably the crossroads of at least two trans-Sahara trading routes. These forts were built in a hurry by the legionnaires themselves, or by local labour, under the supervision of an infantry officer who had received basic engineering training. Even if a regular engineer officer were sent, he only followed the general guidelines he had been given; the rest was up to his own assessment of the local conditions.

The forts were about 25 m (80 ft) square, and when complete were reminiscent of a medieval castle keep with one, two, three, or four corner towers. They were hardly higher than the curtain walls; loftier altitude would have required more substantial foundations. One tower was always used as a sentry lookout and combat command centre. Here waved the

SHURI CASTLE

THE FORTRESS-PALACE of the Kings of Okinawa, built in the twelfth century AD, Shuri Castle's main buildings were of wood. The outlying walls were of limestone and coral, with impressive and ornamental gateways. Shuri's most significant moment was dawn on Monday, 6 June, 1853, when Commodore Matthew Perry USN marched his marines, bluejackets and cannon up from his ships in Naha Harbour and threatened to blast the castle to bits if King Sho-Tai did not open his doors and lands to outsiders. It was the beginning of Japan's Westernization.

Shuri was the Japanese headquarters in World War II, providing an all-round view and with caves below – a feature of the terrain. All attackers were eliminated with flamethrowers and explosives. Offshore, kamikazes attacked Allied warships. Between 1 April and 22 June 1945, 12,500 Americans, 12,000 Japanese sailors and airmen, 108,000 Japanese soldiers, and over 150,000 Japanese civilians died on and around this one island. Now gardens again flourish on rebuilt Shuri.

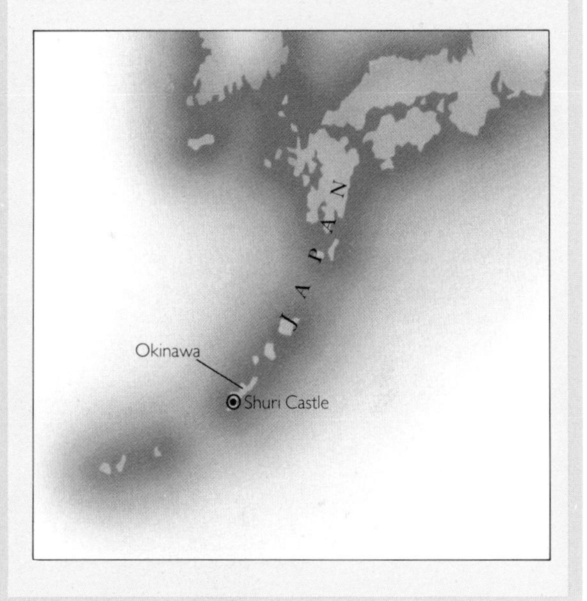

8 July 1824: using scaling-ladders, British troops storm a stockade at Rangoon in Burma. Suffering casualties, they scramble down the defenders' perches, cross an unexpected ditch on the *inside* of the position, and engage in a close-quarters mêlée.

Tricouleur, its very existence heartening Frenchmen far from home, and reminding local people of just who ruled them.

Some forts were surrounded by a dry ditch crossed by a drawbridge which, when raised, added extra strength to the door of the main gate. The approach could be further screened by a defensive earthwork.

The curtain walls of Foreign Legion desert forts were of rubble reinforced with tree-trunks (if available). This core was coated with mud which – once dried – was given a layer of whitewashed lime plaster to repel moisture (such as dew at night), and to keep the interior cool. For further habitability in such a hot climate, the living accommodation was made as open-plan as possible, while the walls and ceilings were only thin lath-and-plaster or wattle-and-daub. Flowers and trees were planted inside the open courtyard to make the harsh environment a little easier on the eye. The barrack rooms, offices and stores backed up against the reverse side of the curtain walls in such a way that their roofs merged with the walkway (or allure) inside the parapet. About ½ m (19 in) thick, the parapet was pierced with rifle slits instead of being crenellated with merlons and embrasures. The latter were not necessary because the walls and roofs were simply not solid enough to bear the weight of cannon (even if it had been possible to move artillery across the desert). Certainly the Arabs did not possess ordnance to any significant extent; nor were they organized or equipped for a lengthy siege.

Desert warfare was a matter of fast-moving, hit-and-run raids. If a tribe could not secure possession of a water-hole or other prestigious property at the first rush, they usually made for an easier target before their own water ran out and their camels died. The besieged, on the other hand, either had easy access to the oasis, or had a well, storage cisterns, or guttering and tanks to collect that rain which did occasionally fall. If they had enough rifles and ammunition to repel the initial assault, they had won that skirmish.

It is probably no exaggeration in fictional films that when rebel tribesmen are seen to acquire artillery from somewhere, each cannonball or shell they fire produces spectacular destruction within the fort. Such places were not built to withstand punishment of this sort.

Conversely, elsewhere in the world, more enduring structures – such as farms, warehouses and churches – were built with extremely solid walls and small windows to withstand the thrust of grain in bulk; with thick foundations to support tall belfries; or with outlying courtyards and walls for keeping in herds of stock. Such buildings often proved formidable defensive structures in an emergency.

Some residences were deliberately built with defence in mind. The Castle of San George d'Elmina was built by the Portuguese soon after their occupation of the Gold Coast in 1481-82. The arcaded gallery enabled the governor's household to enjoy the cool breeze and shade. In a crisis, defenders on that verandah could fire over the heads of the artillerymen on the lower wall and over infantry at ground-level rifle-slits. The whole system was a development of the medieval concentric castle. A counterweighted drawbridge completed the defensive arrangements on that particular side of the fort.

In 1642 the Portuguese were ousted from the Gold Coast by the Dutch. Their rivalry with the English for its possession continued until the Gold Coast became part of the British Empire. By then the slave trade had been abolished and the cells in Elmina Castle, in which slaves had once been kept awaiting export, were empty. But whoever ruled and however enlightened their policy, the whitewalled Castle of Elmina remained on its hill, a perpetual reminder to the local inhabitants that they were under foreign domination.

That was how 12,000 Ashanti warriors regarded it when they crossed the Pra River in June 1873. A few bluejackets and marines were hurried ashore to reinforce the small West Indian Regiment garrison. Their prospects were bleak. However, Elmina Castle had been well designed and well built – and now it was valiantly defended. Though equally valiant in attack, the Ashantis could not storm the place and, lacking siege equipment, in the end they returned to their homeland having demonstrated that they were not a people to be trifled with. They had also unknowingly stirred up Sir Garnet Wolseley, who was despatched to lead a punitive expedition against them.

Of course, as European nations established great empires, their military architecture – even overseas – had to be capable of withstanding the latest Western weaponry their rivals could bring to bear. Eventually their foreign works became as sophisticated as anything in Europe itself, although there was always improvisation in the wilder places on the frontier.

JHANSI

A FORTIFIED city in India 350 km (250 miles) south of Delhi; Jhansi's Rani in 1857 was Tantia Topi, sometimes called Lakshmi Bai after the Hindu goddess of good fortune and beauty. Although not an instigator of the Sepoys' Mutiny, she espoused the cause of Indian independence in June 1857 and organized the defence of her city. General Sir Hugh Rose arrived to besiege it in March 1858 and stormed Jhansi on 2 April, 1858. But by then Tantia Topi had moved her surviving forces north to Gwalior, a typical Indian mountaintop fortress 300 km (200 miles) south of Delhi.

Gwalior's curtain wall extends the full 10-km (6-mile) length of a precipitous ridge with many salients and re-entrants. Within, a Mahratta keep dates from 1754. There is the Painted Palace of Man Singh (1490) and even earlier rock carvings and temples, the latter built about 1093. The main approach is up a very steep track and through a bastioned outwork 1.5 km (1 mile) in perimeter, with complex barbican and gates. It too contains a palace (the Gujari Mahal) and once had a special Wind Gate, or Hawa, which blew a cooling breeze on travellers toiling up the cliff path.

In June, 1858, it was while trying to reorganize her forces at the foot of the crag during retreat after the last British cavalry charge that Tantia Topi, Rani of Jhansi, was sabred. Her body was later recovered by her followers to be buried secretly.

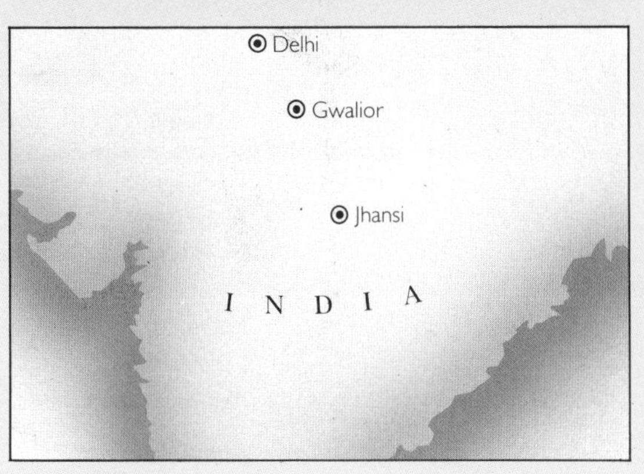

The fort at Jhansi after its capture in 1858 during the Indian Mutiny. Note hilltop location, several parallel walls or switch-lines on lower slopes, the bastions mainly rounded, the pointed merlons, the citadel on the highest part of the mount.

Chapter Five

THE INDUSTRIAL WORLD

The Industrial Revolution began in Britain in the later years of the eighteenth century. The harnessing of steam-power and its application to the mechanical processes of manufacture resulted in hitherto unparalleled output of all types of weapons and equipment during the Napoleonic Wars. Thousands of cannon, muskets, uniforms, pulley-blocks and barrels of gunpowder enabled the Royal Navy to keep the seas in all weathers, continually practising tactics and exercising the great guns, and the British army was able to fight campaigns in every part of the globe. There was still not enough – there never is – and the supply system often broke down, but, overall, the British armed forces and their allies were much better equipped than their opponents.

THE MECHANIZATION OF WAR

Nevertheless, during the long years of conflict British factories were simply mass-producing existing items of war materiel of proven efficiency and quality. There was no time for experimentation. It was not until peacetime that the more technologically-minded officers could find time to investigate the possibility of new types of weapons and equipment – and to argue for their introduction.

They immediately encountered entrenched opposition. How dare anybody criticise the Wooden Walls, the Carronades, the Brown Bess Muskets, and the Martello Towers? They had all beaten Boney – why throw them away for a new-fangled gadget which probably wouldn't work when you wanted it to? In any case, there were enormous stockpiles of weapons, while existing fortifications were good for many years yet. Nobody could justify the expense of making new.

But as the nineteenth century wore on, the Industrial Revolution began to develop its own momentum, especially after the widespread introduction of the machine-manufacture of machine-tools. The competition of free trade was a forcing-house for the development of new processes in mechanics, metallurgy, chemistry, textiles, food-processing, communications, and the transmission of power. Many of these processes had warlike spin-offs, and indeed some industrialists deliberately sought out military markets for their

Above (top): A drawing of 29 July 1871 shows the armament being installed in Fort Gilkicker at Gosport, part of the western defences of the Royal Navy's base at Portsmouth.
Above (bottom): Practice shoot by heavy coast-defence batteries manned by German naval personnel on the island-fortress of Heligoland.

The industrial world on the battlefield: construction of a Federal sap towards the Confederate entrenchments at Vicksburg on the Mississippi during The War Between The States.

products, making contacts among those technologically-minded officers who were now reaching rank high enough to authorize the purchase of new equipment. Inevitably these former 'Young Turks' became dedicated champions of *their* favourite weapons systems, resisting any change until they in their turn were supplanted by newcomers up the ladder of promotion, or until some international crisis precipitated a sudden spurt in the evolution of military technology.

In 1815, smooth-bore muzzle-loading cannon could fire a solid ball weighing 14.5 kg (32 lb) over a distance of 2,745 m (3,000 yd), or one weighing 30.8 kg (68 lb) over a range of 366 m (400 yd). By 1914, rifled breech-loading ordnance could hurl explosive shells weighing 879 kg (1,938 lb) over a range of 21.4 km (13¼ miles) – and with a potentially greater accuracy than its predecessors of a century earlier. Small arms had developed from the smooth-bore musket, firing a single ball every half a minute, to rifled fully-automatic machine-guns spitting out bullets at the rate of 600 a minute.

One result of this greater technicality was the erection of military workshops, training centres, lecture halls, barracks, hospitals and a whole range of support structures hardly different from civilian buildings of similar function. However, the most obvious effect of the Industrial Revolution on military architecture was in the design of military fortification. The problem was twofold: how to keep out the projectiles launched by the new weaponry; and how to install and operate that same equipment. The devices and stratagems became incredibly complex, each new weapons system worthy of a complete study in itself.

The end result was that by the beginning of the twentieth century the new fortresses had become artil-

SEVASTOPOL

DURING THE reign of Catherine the Great, the Crimea was captured from the Turks and the Russian Navy was revitalized (partly by the American John Paul Jones). Sevastopol was therefore founded in 1784 as the main base of the expanded Black Sea Fleet. Its sea-facing forts – updated with shell-firing armament – dominated deep-water channels through the shallows. During the Crimean War they prevented assault by the Royal (British) and French Navies. Meanwhile the land defences (organized by the engineer General Todleben) stood equally firm against the combined armies. Much of the fortification on both sides was in the nature of fieldworks, some being so novel that that particular part of the battlefield acquired the name 'the Rifle Pits'. However, the Russian lines had two principal purpose-built strongpoints: the Great Redan and the Malakov. They withstood bombardment and assault throughout the winter and summer from 28 September, 1854 to 8 September, 1855, when the Malakov was stormed by the French, and the rest of Sevastopol's defences became untenable.

During World War II, the German army broke into the Crimea on 26 October, 1941. By 16 November, 1941, Sevastopol was the only Russian enclave on the peninsula. The Germans could not ignore it because of the dual threat its 106,000-strong garrison (men and women) and naval forces posed to their land and sea communications. The effort and delay in capturing Sevastopol by midsummer 1942, however, probably cost them Stalingrad.

lery strongholds, bases of operations – not for columns of infantry or cavalry, but for heavy shells whose targeted delivery could destroy crossroads, bridges and railways over a radius of a dozen km (8 miles) or more. The targets did not even have to be within sight of the fort. Thanks to accurately-printed maps, fire-control instruments, and electrical communications, neither intervening hills nor the hours of darkness prevented a fort of 1914 from completely dominating the surrounding countryside – at least in theory. Properly sited, each fort helped to cover its neighbour with overlapping fields of fire. A ring of such forts could defend a city; a line of them could defend a frontier. Defence in depth, made up of a series of parallel lines of fortresses, strengthened by fortified complexes around manufacturing centres, could turn the whole state into one titanic, impregnable stronghold.

This was the reasoning behind the French, Prussian and Dutch constructions of the nineteenth century. Indeed, every European power subscribed to this strategic thinking in some way. Forts, battleships, size of army, and railway mileage, were all symbols of national status.

Left: It would seem that the pre-Industrial Revolution artist of this view of the 1779 siege of Gibraltar saw neither the place nor a map of it. Hence his fanciful castle surmounting the Rock and other romantic imagery.

Above: Not all Maginot Line fortifications were purpose-built in the 1930s. La Citadelle de Bitche (Moselle) saw service in both 1870-71 and 1914-18. It was later reconstructed as a strong-point slightly to the rear of the Line itself.

However, the ability of a field army to win wars – or at least avoid losing them – was what still determined the outcome of a conflict, no matter how effective the static fortifications or spirited the defenders. That was one of the lessons of the Franco-Prussian War of 1870-1871. The Near East Crisis three decades earlier had persuaded the French government to do something about the absence of defence works around Paris – a deficiency which Napoleon Bonaparte had belatedly regretted even earlier. Accordingly, a building programme was initiated which, by 1870, had surrounded the entire city with a moated wall 10 m (33 ft) high and equipped with 94 bastions mounting a total of 800 heavy guns. Another 700 heavy guns were held in reserve. Beyond The Enceinte were 15 detached forts, each armed with 100 heavy guns.

It was in July, 1870 that a diplomatic dispute over the succession to the Spanish throne brought about a French declaration of war against Prussia. By September the Prussians had invaded through Alsace-Lorraine, defeated the main French field army, and were approaching Paris. A Provisional Government of National Defence improvised fieldworks, blocked roads and railways into the capital, installed mines and booby-traps, and cleared fire-zones in front of these barriers. Forts denuded of garrisons were manned by sailors transferred from the fleet . . . all to no avail. The Prussians reduced the detached forts with longer-ranged field artillery and then stormed in. The defenders were reduced to holding out in little mounds of gabions and brushwood fascines. The stranglehold on Paris was complete, while the Prussian field army held off French forces trying to break through from the Loire, and ruthlessly exterminated *franc-tireurs* (civilian snipers). The siege of Paris lasted from 19 September, 1870 to 28 January, 1871; the city's surrender marked the establishment of the German Empire.

But although the permanent fortifications around Paris had not saved France, they seemed to vindicate the proponents of fortress construction. After all, the citizens of Paris had held out for four winter months – and had still had sufficient energy subsequently to rebel against an unwanted government. Certainly the newly-created state of Belgium, its neutrality guaranteed by all those nations likely to invade it, saw no reason to halt the programme of fortification begun in 1859 under the aegis of Henri Brialmont.

The whole of Antwerp was enclosed by a linked

AMERICAN COASTAL DEFENCE

Masterpieces of fortification which allowed small numbers of troops to defeat the onslaughts of the British fleet

Once the colonies in North America had passed beyond the pioneering stage and become established political and economic duplicates of their homelands, it became clear that the most direct threat to their existence was posed by transatlantic amphibious assault by rival European powers. They therefore began the construction of forts denying unauthorized naval access to strategic centres.

To a certain extent the theory behind it was an adaptation of the traditional British strategy of fleet-forts-field army. In North America, however, the targets were so widely scattered and so separated by mountain, marsh, forest or water, that troops had usually to be transported all the way to their objective. American coastal defences therefore laid greater emphasis on artillery fire and protection against bombardment, rather than on the provision of bases for the repulse of infantry assault, or of barracks for the accommodation of reserves for mobile land operations.

Forts had to be able to withstand a siege, but such an ordeal was likely to be of shorter duration than in Europe because of the difficulties in maintaining transatlantic supplies and because of the weather – cold and ice in winter, and the tail-end of hurricanes in late summer.

Prolonged close investment of coastal defences was thus rare. Naval operations were more often some sort of blockade, with the possibility of occasional runners getting through to beleaguered garrisons. Attempts to deny potential bases to the enemy meant that sometimes forts (which may subsequently have expanded into colonial settlements) were created simply to establish right of occupation.

The need to prevent interference with colonial trade meant that

Above: 28 June 1776: cannonballs from British warships bombarding Sullivan's Island Fort brought down the American flag. For a moment, the watchers in Charleston thought that the defenders had surrendered. But "Sergeant Jasper jumped from one of the embrasures and brought it through a heavy fire, fixed it upon a sponge-staff and planted it upon the ramparts again, and revived the drooping spirits of our friends."

fleet commanders often had to confront the enemy far out to sea, a necessity that in turn laid greater emphasis on the provision of dockyards for refit and repair, and on the defence of these vulnerable targets. It also meant that these fortifications tended to reflect European trends in construction and armament more readily than frontier forts.

In 1613, for example, the French settlement of Port Royal (founded in 1605 on the south shore of the Bay of Fundy) comprised a simple ditch of square plan, within which were several houses. The only form of fortification was a lozenge-shaped earth bastion facing the river. By 1751, (by which time it had been captured three times by New England colonists and was now permanently British), a star-shaped fortress crowned the hill overlooking the town and dockyard. Both the latter were further protected by a bastioned wall along the waterfront with square blockhouses at particular danger-points. The place had in addition been renamed Anna-Polis in honour of the Queen.

An engraving by the American silversmith Paul Revere shows North Battery at Boston in 1770 to be a stone-walled fort rising from the water. Its face is concave to withstand breaking waves, and its corners are embellished with two loopholed bartizan-towers. Between them, at least 13 cannon can be seen

Left: The angled bastions of Fort George in New York (1760) reflect European trends in military architecture. However, in the event of hostilities it would be necessary to fell the decorative trees between the fort proper and the outer entrenchments and batteries.

Below: Symbol of Washington's government in seceded South Carolina, Fort Sumter is bombarded by Confederate artillery in April 1861. In spite of the violence of the cannonade, none of the garrison was injured.

in embrasures just below the parapet. Over the fort flies a huge Union flag, symbol of British dominion over the colony.

Although there is no evidence other than this drawing, it can be assumed that Boston North Battery was of stone – earth would quickly have been washed away by the sea. Many southern colonial forts, however – like the one on Sullivan's Island, guarding the approaches to Charlesto(w)n, South Carolina – were of palmetto and earth (very resilient tree-trunks covered with impact-absorbing clay). The parapet was 5 m (16 ft) thick and high enough to give protection from marksmen in the fighting-tops of ships.

The fort was still under construction in June 1776 when the Royal Navy arrived to put down rebellion in this area. Troops were landed to search for fordable shallows while nine men o'war took up bombardment positions only 800 m (880 yd) from the fort. They brought a total of 260 guns into action; Sullivan's Island fort could only reply with 26. Nevertheless, after 10 hours' action the badly mauled British fleet withdrew, apart from one frigate aground and later burned. The British had suffered 106 killed and 65 wounded; the Americans 12 killed and 24 wounded, mostly by shot coming through the embrasures.

The British forces did better at New York in 1776, even though the fortifications there were more extensive. Some had been improvised, whereas others were of earlier, permanent construction, including what became known for ever as 'The Battery' at sea-level on the south-western tip of Manhattan Island. The Narrows defences, however, were not strong enough, and the Anglo-Hessian army reached the shore. They marched through difficult terrain and outflanked the American fieldworks, defeated the outnumbered and less-equipped colonial troops, and forced them to retreat across the East River. The British followed up their advantage and won New York.

Above: Market Battery and Shoal (or Victoria) Tower at Kingston, Lake Ontario, begun in the winter of 1847. Note the conical roof to protect the armament from snow.
Below: Corregidor Island (Fort Mills) in the Philippines. Three of the four M-1890 12-inch (305-mm) mortars, the nearest of which remained in action until the very last day of the siege in May 1942.

Not always so successful, the British did, however, retain considerable territory in North America and the Caribbean, which now not only had to be defended against European rivals but also against the possible spread of revolution from the former colonies. Among the naval bases whose defences were accordingly strengthened was Nassau in the Bahamas.

Fort Fincastle, built in 1787, is a circular building 26.7 m (84 ft) in diameter and 2.8 m (9 ft) high, mounting three heavy cannon, which fired over the parapet and across the harbour. Four m (12 ft) above the cannon, two smaller guns fired through embrasures against infantry and artillery attacking overland. This was presumably seen as the approach zone of greatest danger, for between the two guns a huge sloping wedge runs to a point 25 m (80 ft) away. It is obviously intended to deflect cannonballs and mortar shells from reaching the magazine and the other two rooms hollowed out of the solid mass of masonry. It looks just like the bow of a future ironclad battleship, built in stone. Its distinctive appearance, so comparatively close to the mainland of North America, may well have influenced US military architects in a later century.

A definite line of development

Above: The unusual shape of Fort Fincastle at Nassau in the Bahamas.

Below: Fort Drum after the American return to the Philippines. The Japanese still inside refused to surrender, so on 13 April 1945, US Army engineers boarded 'The Concrete Battleship' and poured in fuel to burn and suffocate the last survivors.

from British designs were the Martello Towers built at Halifax (Nova Scotia), Quebec on the St Lawrence, Saint John (New Brunswick), Kingston (Lake Ontario), Charleston (South Carolina), Tybee Island (Georgia), Lake Borgne (Louisiana) and Key West (Florida). They all varied in type of location – one was erected on a cofferdam sunk in the water – and in detail. Some were hexagonal, some had hemispherical caponiers, and at least one had machicolation, quite apart from the ones of traditional appearance. Some were built of tabby – a locally available, but easily erodable, mixture of seashells and lime. Some performed other roles (one was a lighthouse) and all the surviving ones were utilized as World War II lookout, signalling, radio or anti-aircraft positions. They all owe their origin to the War of 1812 between Britain and America, and the period of uneasy peace that followed.

The garrison at Fort Sumter in Charleston Harbor, South Carolina, totalled a mere 84, and it was rationed for only a few days, but it was a Federal enclave in seceded Confederate territory and could be supplied only by sea. At first, relieving ships were fired on until they withdrew, their mission unfulfilled. Then the fort itself was bombarded by 50 cannon from 12 April 1861 until it surrendered on 14 April 1861. And so began The War Between the States.

During the next century, American coastal defence reflected not only the growing range of artillery in particular, and of weaponry in general, but also the expanding interests of the United States. These were especially vulnerable in the Far East. Long-range defence against invasion of the Philippines was provided by the US Navy, operating from its base at Cavite. Security there was guaranteed by a series of fortress-islands in Manila Bay, completed in 1914. Corregidor was the biggest, and had 56 mortars and guns (including disappearing mountings), a barracks and (later) an underground bomb-proof complex of accommodation, offices, hospital and stores. The most unorthodox was Fort Drum.

From 1908, engineers blasted away the upper surface of El Fraile Island and honeycombed its interior with tunnels, stores and accommodation spaces for 200 men. They then smoothed its top and sides, covering the entire island with a concrete shell 7.5-11 m (25-36 ft) thick, 106 m (350 ft) long, 44 m (144 ft) wide and 12 m (40 ft) high. The armament comprised four 15.2-cm (6-in) guns in naval-style casemates at the side, and four 35.6-cm (14-in) guns with a range of 30 km (18¾ miles) in two massive super-firing naval-style turrets. Fire-control and signalling was conducted from the top of a basket mast

exactly like those installed in US battleships of the period. With a sharp end like a bow and a blunt end like a stern, Fort Drum looked so much like a warship that it was often called the 'The Concrete Battleship'.

Unfortunately, when war came in December 1941, Japanese troops occupied the mainland while their air force destroyed the naval base. Without a fleet and field army, the forts were forced to surrender on 6 May 1942.

system of 11 lengths of fortification. The northern end formed a citadel with threefold function: it mounted a coastal-defence battery overlooking the Scheldt estuary; it served as headquarters; and it was the defenders' last refuge if the rest of Antwerp and Belgium had been overrun. Then came a ring of 18 forts, redoubts and earthworks. 11 km (7½ miles) beyond that came an outer ring of 31 detached fortifications. Possible German passage through the Meuse valley was blocked by 21 forts ringing Namur and Liège. The largest forts could accommodate 1,000 men, and 135 guns and mortars up to 21 cm (8.3 in) in calibre. If all were fully equipped, some 9,000 pieces of artillery were required, their operation needing the attention of 70,000 men, quite apart from the 100,000 in the Antwerp garrison itself, plus however many were to make up the Belgian field army.

Transcontinental countries like Russia, the United States and the worldwide British Empire could hardly hope to defend their extensive frontiers with continuous lines of fortification. Nevertheless, they followed the same policies of creating national strongholds and building fortresses to block the strategic approaches to vital areas of territory. Usually these took the form of coastal defences, protecting the vital dockyards and bases for those battle-fleets whose success or failure would be the deciding factors in global war.

In 1863, for example, the British built forts in the sea off Portsmouth. Civil engineers had developed the use of cofferdams (walls of piling, forming watertight enclosures from within which the water could be pumped out) and caissons (huge cylinders containing air at high pressure to keep the water out) sunk into the seabed. Both systems enabled labourers to dig out the subsoil and lay foundations in a comparatively dry environment. The seafort foundations were of granite and Portland stone, forming a ring wall 16.3 m (53½ ft) thick and 70 m (230 ft) in diameter, enclosing a massive disc of cement-covered rubble. (Alternative methods were to lay the stones directly on the seabed, employing divers to position and key them; or, even more quickly, to tip in blocks of stone or concrete until they rose high enough to form an artificial island. It all depended on the condition of the seabed, currents, and climatological factors.)

Once the base was complete, a circular stone fort

The mid-nineteenth-century sea-fort just inside the breakwater at the entrance to Plymouth Sound. The passages round each end of the breakwater were dominated by shore batteries at Cawsand in Cornwall, (on the right) and Bovisand (behind the camera).

Left: Shargai Fort built by the nineteenth-century British Army on India's North-West Frontier (now Pakistan). Serving as defence against invasion through the Khyber Pass, and as a base for operations into Afghanistan, it had to combine defensive features and accommodation for personnel, horses and stores. It also required areas for training, relaxation and tattoos (peaceful displays of military power). Note the names of Fort and Regiment picked out in white stones.

Right: Representing two constituents of Britain's fleet-forts-field army strategy, an 87-foot (36.5-metre) Thornycroft 1st Class torpedo-boat dashes around one of the Spithead sea-forts in 1888.

was constructed capable of mounting 25 10-inch (25.4-cm) 18-ton and 24 12.5-inch (31.8-cm) 38-ton rifled muzzle-loading (RML) guns, in two tiers. This armament could deliver explosive shells totalling 13.4 tonnes and capable of smashing through 30.5-45.7 cm (12-18 in) of iron plate at about 1,000 m (1,100 yd) range. The forts themselves were protected from horizontal fire by three layers of iron armour-plate, varying in thickness from 4.6 cm-6.4 cm (1¾-2½ in), separated by two layers of iron concrete (a mixture of asphalt and iron filings or swarf). This energy-absorbing armour sandwich was bolted through an inner layer of wood and concrete on to the walls of the fort. The latter supported the iron gundecks without being fixed to them. The walls could thus be battered by enemy shot without the shock jamming the guns. And, just in case the enemy should bring up high-trajectory bomb-ships, the roof of the Spithead forts was made up of concrete 1.4 m (4½ ft) thick.

This type of seafort was also built off Royal Navy bases at Portland, Plymouth and Bermuda. They were not all the same size, nor of exactly identical design. Some were planned to mount revolving turrets on the top; others dispensed with part of the vertical armour-plate if shallow water prevented hostile warships from passing on one particular side.

These forts were virtually stationary ironclads, but more conventional fortifications were simultaneously under construction ashore. In particular, there was a whole series of polygonal forts on Portsdown Hill to defend Portsmouth dockyard from bombardment by French troops who – it was feared – might come ashore in Sussex to swing round overland instead of trying to force their way in through the harbour entrance. Features of these forts included brick arches built into the masonry for reinforcement against bombs falling from high-trajectory mortars. Double-storey caponiers jutted into the dry ditch, while an occasional sally-port half-way up the fortress wall enabled the defenders to deal with any enemy who survived his fall into the ditch. (The enemy would be unable to get at the sally-port door himself because of grapeshot and rifle fire from the caponier.) Everything was done to delay hostile control of Portsmouth; every minute gained time for

HELIGOLAND

'The sea-girt fortress', its beetling cliffs bristling with huge gun emplacements, its interior honeycombed with barracks, magazines and titanic caverns all ready to disgorge battleships, U-boats and sea-plane-carrying zeppelins upon unprovoked and undeclared warfare against the shores of England was Heligoland as portrayed by Percy F. Westerman and other novelists between 1910 and 1914. The reality was less dramatic and more defensive.

A small island 52 km (33 miles) from Germany, it was exchanged by Britain for German territory in Africa in 1894. The Germans then fortified it so that its artillery covered German warships exercising off their main fleet bases, and updated the fortifications both before and during World War II, including the installation of formidable anti-aircraft defences. It was bombed on a number of occasions, but usually as a target of opportunity when a formation could not find a strategic or more significant tactical objective.

Towards the end of the war Heligoland could be ignored no longer – diehard Nazis might hold out in its fortifications, preventing Allied naval movements in an otherwise peaceful Europe. So on 18 April, 1945 Heligoland was devastated by 969 aircraft of RAF Bomber Command. In 1946-1947 all Heligoland's surviving defences were blown-up by Royal Engineers.

Above: This World War I view of Heligoland epitomizes the sinister reputation which Germany's island-fortress had amongst British navalmen and aviators. It seems that the gunsite seen on page 130 is under construction.

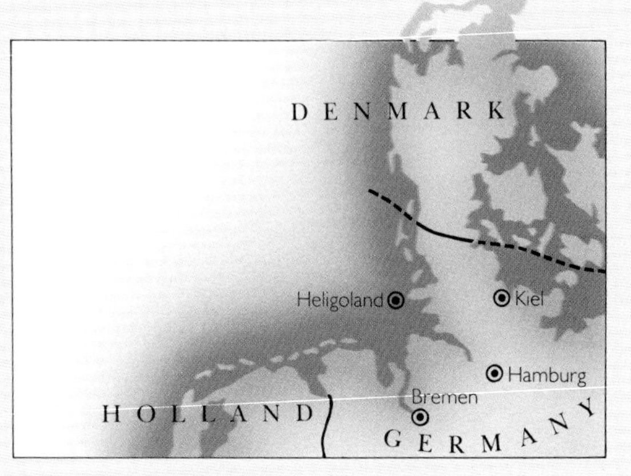

Above: The location of German Heligoland in the North Sea.

Below: The end of the guns seen on page 130.

Below: German naval artillerymen loading a shell-hoist beneath one of the heavy guns seen firing on page 130. Although lit by electricity, paraffin battle-lanterns hang ready in case of power failure.

Above: Fort Siloso, Singapore. One of the 6-inch (152-mm) gun emplacements with ranging tower, magazine, and other buildings.

the fleet to put to sea. Once beyond range of enemy artillery, there would be no doubt of Royal Navy victory over the warships and transports upon which the invading French land army would have to depend. Combined with the construction of new battleships like HMS *Warrior*, and the establishment of the Volunteer Movement ashore, it was a classic example of the fleet-forts-field army strategy.

In the event, the French never came – which caused the Portsdown Forts to be nicknamed 'Palmerston's Follies' after the Prime Minister of the day. The deterrent effect of such fortifications was completely ignored by his critics. In any case, they argued, seaport defences ought always to face the ocean. (Exactly the *opposite* argument was used in hindsight after the Japanese had captured the naval base of Singapore by overland assault in 1941-1942.)

One of the biggest problems faced by military architects resulted from the need to store ammunition as safely as possible – ie as deep underground as possible – while mounting guns as high up as possible. The development of power-driven hoists helped, but the time of greatest danger and complication was still during reloading, which always took longer than the actual firing. This was especially true of rifled muzzle-loaders – the entire piece had to be run back in and depressed below parapet level.

One answer was the disappearing mounting, in which the gun popped up, banged away, and disappeared again. It certainly presented a very small target to the enemy, and was very spectacular. It could be said to have been the heavy artillery equivalent of the quick-draw gunfighter of the Wild West – and it was probably as accurate!

Accurate or not, the disappearing mounting was only protected frontally; it was still vulnerable to plunging fire from mortars and from the longer-range higher-trajectory artillery coming into service by the 1880s. Such protection may have been adequate for coastal defence, the attackers only coming from one direction – but it was a different matter for artillerymen in land forts likely to be outflanked. So Lieutenant-Colonel Schumann, a Prussian, proposed covering the disappearing mounting with an armoured roof – a cupola – that went up and down with the piece, completely shielding the whole mechanism when retracted.

SINGAPORE

As envisaged in 1921, Singapore was a *naval* base, its warships intended to destroy the enemy long before they came within range of the coastal defence guns. These themselves were quite impressive; many of them could be traversed to cover the land approaches to the island. Troops and aircraft were allocated to the area, but they were principally for policing the mainland and could not really be considered a field army. But that did not significantly matter, as France occupied Indo-China, the Netherlands occupied the Indies, and there was a friendly government in Siam (Thailand).

But by 11 December, 1941 the Japanese were in **Indo-China,**Thailand and Malaya, the US Navy had been eliminated, and Singapore's own battlefleet annihilated. The classic historical British strategic policy had proved to be fleet-forts-field army; without one element, there would be difficulties; without two, there was defeat – the surrender of Singapore occurred on 15 February, 1942.

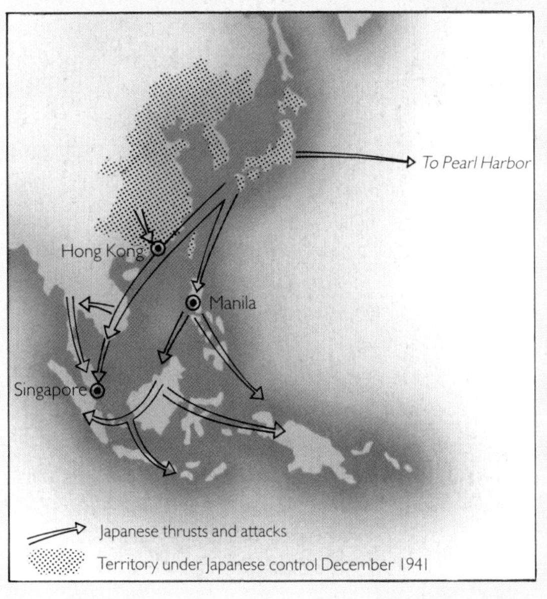

Above: Close-up of the 6-inch (152-mm) emplacement, ranging tower and (in the background) the magazine entrance at Fort Siloso on the north-west tip of Sentosa Island (originally called Pulau Blakang Mali Island) in Singapore. The gun is actually a 9.2-inch (235-mm) piece from Fort Connaught (also on Sentosa Island). It dates from World War I and had been so sabotaged by its British artillerymen in the last moments of the siege that the occupying Japanese made no attempt to use it. In any case, there was no ammunition left out of the thirty rounds supplied for each gun! The Japanese never got around to cutting up and melting down this particular barrel as they did most of the others.

Left: Singapore's strategic location – a strategic trap when the surrounding Indies were occupied by a hostile power!

Monsieur Mougin of the French Saint-Chamond munitions factory was also a contemporary advocate of cupolas, as was Hermann Gruson of Krupp's. Umbrella-shaped cupolas hiding vertically-retractable turrets became all the fashion in military architecture.

And with the reintroduction of breech-loading ordnance, cupolas did not even have to be retractable. In some types, the muzzle of the gun projected from something like an upturned metal saucer, which had all-round traverse and as high an elevation as existing mountings would allow. No matter how much it moved in any direction, the rim of the saucer was always concealed beneath an apron or glacis of concrete 2.5 m (8½ ft) thick, most of which was further shielded by three m (10 ft) of earth.

Indeed, the magazines, stores, workshops, and accommodation were all cut out of the solid rock below and connected by tunnels. All was underground, illuminated by electricity produced by internal combustion engines, electricity which also powered all the machinery necessary for living in and operating the fort. Once the scars of construction had been concealed by grass and a few bushes, the only visible items of military hardware were the low cupolas, an occasional ventilator or rifle-port, and a solid door for infantry counter-attack in an artificial-looking squared-off dry ditch. Only if a hit were scored right on one of these very small targets could any damage be done.

Secure behind their frontier fortifications, the industrialized nations built up the rest of their war machines, equipping field armies and base complexes. The accumulation of provisions, munitions and reinforcements – once confined within the dungeons and rooms of individual medieval castles – was now spread throughout the land, transforming each state into one vast stronghold. The industrial world had prepared itself for two global wars and the spawning of even longer-lasting conflicts.

U-boats fitting out at the Germania Yard, Kiel, owned by Krupp, whose steelworks produced so much war material (including fortifications) that the name became synonymous with German armaments.

THE MAGINOT LINE

The defensive system to beat all systems was rendered impotent by the strategic skills of the German Army

La Ligne de Maginot was intended to defend France with masonry instead of with blood. The project was suggested by Marshal Pétain (defender of Verdun) in December, 1925, but its chief proponent was André Maginot, appointed French Minister of War in November, 1929.

The Maginot line was made up of a series of *ouvrages*, or 'works'. At the heart of each was a maze of tunnels and chambers excavated out of the solid rock and lined with cement. Here, safely buried beyond the deepest penetration of the biggest shell or bomb were the magazines and stores, power-stations, battle-headquarters, hospitals and barracks for up to 1,200 men, equipped with every facility for troop efficiency and welfare. Far above, fed by ammunition hoists and controlled from separate command posts receiving reports from observation bunkers, were the steel cupolas, both fixed and disappearing types, mounted in reinforced concrete glacis. The biggest were 4 m (13 ft) in diameter, mounting a 75-mm (3-in) gun with a range of 12 km (7½ miles). There were also 135-mm (5.3-in) mortars delivering a 21-kg (45-lb) bomb over a distance of 5.6 km (3½ miles), plus a whole variety of lesser weapons.

The largest *ouvrages* were in two blocks – infantry and artillery – which were armed with lighter and heavier calibres respectively. Up to 2.5 km (1½ miles) apart, the two blocks were linked by an underground electric railway, which also ran to the entrances to take person-

nel and materiel back behind the firing-zone. The entrances, too, were well protected by armoured doors, machine-gun slits and false gateways. Inside, at various positions along the tunnels, were ambuscades and hidden mines to entomb unwary invaders, while the defenders had their own secret exits and sallyports. Everywhere there were fireproof and gasproof doors, exhaust vents and air intakes fitted with anti-gas filters.

In front of the *ouvrage* was a line of two-tier casemates, armed with anti-tank guns below, and lighter weapons above, each one a smaller version of the great underground forts. Their visible portions were surrounded by an anti-tank ditch, with vertical posts ('asparagus'), barbed wire, booby-traps and mines forming further obstacles to mechanized and human advance.

About 5 km (3 miles) in front of these casemates were the blockhouses, also of ferro-concrete. Ostensibly forward observation posts for the main line of *ouvrages* and casemates, they were equipped with anti-tank guns and automatic weapons. Each 30-man garrison was capable of putting up a fierce fight on its own.

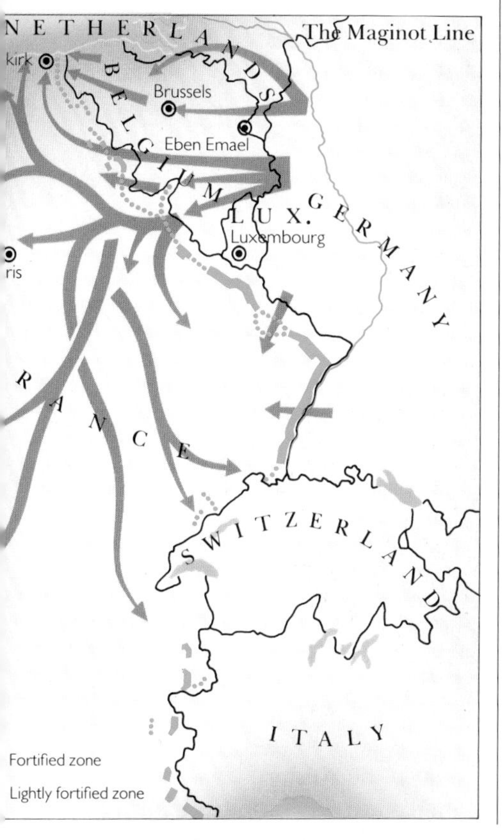

Far left: This is how many people believed the Maginot Line to be, rows of surface cupolas suggesting continuous defence.
Left: The real map reveals that the Line was not continuous. The photograph **above** (of Hackenberg) shows that the interior was spartan and gloomy. Nowadays tourists come to see the working 135-mm (5.3-inch) turret of Hackenberg Block 9 (**below left**). In December 1944, the visitors to Maginot (**below**) were Americans liberating French civilians from Frohmond village, hiding from the retreating Germans.

From here the German border was only 2 km (1¼ miles) away. Every road junction from the blockhouses to the frontier was prepared for demolition; every bridge on the frontier itself was covered by fire from a fortified house. Each fortification was so strong that it could be bombarded by adjacent positions without damage, but with devastating effect on any Germans clambering about outside. And if, in spite of all this, any invaders did manage to run the gauntlet of the Maginot Line, they were then to be mopped up by the French field army, while mobile anti-aircraft batteries and the air force dealt with the enemy in the sky.

The Maginot Line was never intended to be a continuous solid fortification. Indeed, even the belt of defences that was completed was as complex as outlined above in only two sectors: Rhin-Bas and Moselle, and Moselle and Meurthe-et-Moselle. Other sections of the Franco-German border were covered by lesser strongpoints considered sufficient for those more mountainous regions. The Belgian frontier was not fortified at all for a variety of reasons, although they had their own new underground fortress at Eben-Emael, dominating the bridges over the Albert Canal north of Liège.

But when war did come to the Western Front, in May 1940, that was the way the Germans came, through the Ardennes and the Low Countries. The Maginot Line was outflanked and forced to surrender although most *ouvrages* were still holding off assault from both the Front and the rear right up until the Armistice on 22 June, 1940.

OSCARSBORG

JUST HOW effective coastal fortification can be was demonstrated by the fortress of Oscarsborg at Drobak, overlooking the narrowest part of Oslofjord. Although neutral on the night of 8-9 April, 1940, the Norwegians had a minelayer (*Olaf Tryggversson*) patrolling the entrance to the fjord. She gave the alarm and opened 12.7-cm (5-in) fire on an unauthorized force of warships coming by. The German cruiser *Emden* was hit and the torpedo-boat *Albatros* badly damaged and beached. Meanwhile the cruisers *Blücher*, *Lützow* and *Emden*, two more TBs, and 10 minesweepers pressed on up the fjord – perfect targets for the alerted 28-cm (11-in) gunners and shore-mounted torpedo-tubes at Oscarsborg. Within moments, *Blücher* – a brand-new ship the size of HMS *Belfast*, and carrying 1,000 assault troops and their ammunition – had blown up and sunk, and *Lützow* (hit by three 28-cm (11 in) shells) was turning back with the rest. The Germans did capture Oslo later that day, but by airborne invasion, by which time King Haakon VII and the Norwegian gold reserves were on their way to safety. Oscarsborg had not prevented the occupation of Norway, but it had bought enough time to conserve Norway as a political and economic entity.

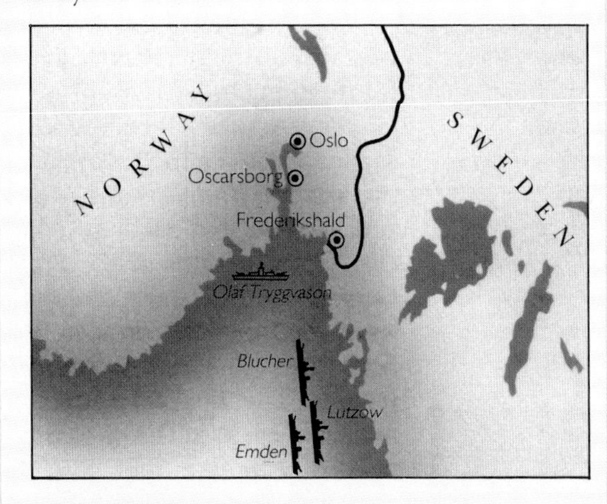

Above: Oscarsborg: in the narrows, Blucher will be sunk; Lutzow and Emden will be damaged.

THE 1914-18 WAR

The events of 1914-1918 and 1939-1945 can be considered one single interrupted period of hostilities which saw warfare transformed from movement to stalemate, and then the rapid thrust-and-parry of armoured and aerial forces. Forts played a significant part in both strategy and tactics, although prejudiced views of their achievements were subsequently voiced both for and against them.

The opening moves on the Western Front involved the Belgian forts. Timing was vital; the German plan depended upon pushing their main striking force through Belgium as part of their opening – and victorious – gambit. It is ironic that this scheme had been devised because of the rival fortifications on both sides of the Franco-German boundary. They were of comparable strength, but whereas the French intended a spirited frontal attack on the German positions and a quick dash to Berlin, the Germans decided to outflank the French positions altogether. The bulk of their army would swing west, south, and east to take the French in the rear, sweep them up, and hammer them against the belt of German forts along the Rhine. To do that the German army had to go through Belgium – preferably peaceably, but if not, then forcibly. And that meant taking out the Liège forts.

Military experts did not think this was feasible. It was true in 1914 that the Liège forts were not of the latest design and were somewhat exposed, being made

Local defence positions at Fort Loucin near Liège in Belgium, as built before World War I. Steps lead from underground shelters to revetted entrenchments, enabling small arms fire to bear on enemy engineers trying to assault the fort's artillery turrets.

up of a central citadel with parapets to repel infantry attack, all set within a dry ditch, triangular in trace and provided with counterscarp galleries. Nevertheless, both armour and construction were believed proof against the 21-cm (8.3-in) field howitzers that were all that could be deployed in mobile warfare at that time. Experiments suggested that a succession of shells could be placed fairly close together from a distance of 900 m (1,000 yd) and thus gradually smash through the surface protection, but the general consensus was that the ideal conditions of undisturbed firing were not likely to be repeated on campaign.

A decade earlier, however, the Japanese had effectively employed 28-cm (11-in) howitzers against the Russian forts at Port Arthur on the Pacific shores of Manchuria. These guns were actually redundant naval equipment, their 318-kg (700-lb) shells fitted with delayed action fuzes so that they could pierce armoured decks before exploding. That was how the Russian battleships still in harbour had been sunk in a single day's bombardment. The reduction of the forts, however, had taken them from 1 October, 1904 to 2 January, 1905, which was generally seen as an indication of the howitzers' lack of success.

For some reason, of all the military observers on both sides, only the Germans fully appreciated the fact that the Japanese shells had actually penetrated the Russian concrete and that subsequent delays in capturing the fortifications had been due to other factors. Accordingly, German artillerymen and industrialists set to work to devise and deliver a secret weapon capable of reducing any fort.

On 2 August, 1914 Germany demanded passage through Belgium. This was refused, and by 4 August, the two countries were at war. Liège itself succumbed to surprise attack on 7 August, 1914.

The other forts around Liège fulfilled their expectations; they stopped the German advance through the Meuse valley. Only at Fort Barchon was the German 34th Division able to exploit dead ground and attempt a direct assault at close range. The defenders rushed out on to the parapets to drive them off, unaware that a battery of 7.7-cm (3-in) and 10.5-cm (4.1-in) howitzers from the German 27th Division had worked its way round on the other side. With shells falling upon their unprotected backs the Fort Barchon garrison surrendered. However, other forts were still holding out, dislocating the precise timetable of the Schlieffen Plan.

But *Big Bertha* was on her way, a 42-cm (16.5-in) howitzer allegedly nicknamed after the wife of Krupp von Bohlen und Essen, the German industrialist. Weighing 42.5 tonnes, it was an unwieldy piece of battlefield artillery, but when dismantled into five great sections towed by huge tractors it was just about mobile.

By the evening of 12 August, 1914, *Big Bertha* was in position and had begun firing 1.15-tonne shells over a range of 9.4 km (6 miles) with great accuracy. The eighth projectile established the range, and by next day another howitzer had arrived. Their shells plunged through earth and concrete, exploding inside the fortifications, stripping away whole sections of vertical retaining wall, blasting away the gun mountings from underneath. Cupolas and turrets were penetrated and undermined, they collapsed and overturned, their exposed mechanisms smashed to pieces; the garrison

Left: Big Bertha's handiwork at Fort Loucin in Belgium, 1914. Note that it is the concrete not the armour which has disintegrated. Note, too, the cupola's exit door – still jammed shut.

Below: Looking along a belt of barbed wire in the Hindenburg Line between Arras and Cambrai. The ruins of Pronville look harmless, but they will undoubtedly conceal a variety of emplacements.

pulverised, entombed, poisoned with fumes and driven mad, was forced to surrender. On 16 August, 1914, the last of the 11 fortresses capitulated, and the German army was stepping out down the roads of Belgium on its way to victory.

There were other fortresses at Namur, Maubeuge and Antwerp, but the *Kurz Marine Kanone Batterie Nr 3* was now up to its full strength of four 42-cm (16.5 in) howitzers. Namur was finished as a viable defence system on 26 August, 1914; Maubeuge followed; and Antwerp's defences were bombarded from 4 to 9 October, 1914. Next day the Germans were in the city.

The German advance was stopped eventually, and by fortifications, but these were in the form of temporary fieldworks consisting of trenches, dugouts and barbed wire. Allied counter-attacks foundered on identical obstacles, and soon parallel continuous lines of entrenchments stretched from Switzerland to the sea.

Yet even though they had been overrun, the Belgian strongholds had ultimately achieved one of the aims of fortification: they had irretrievably delayed the German advance, enabling the British Expeditionary Force to arrive on the Western Front to the discomfiture of the leading German troops – a dislocation that worked to the opportune advantage of the French high command and hence led to the checking of the German onslaught. The Belgian forts had bought time.

An even longer delaying action was fought by the Austro-Hungarian fortress-complex around Przemysl at the head of the San valley leading down into Russian Poland from the Carpathian Mountains. Przemysl prevented the Russians from completely following up any temporary advantage gained during the to-and-fro campaigns of the winter of 1914-15. It is true that its investment from 12 November, 1914 to 22 March, 1915 meant that the Austro-Hungarians could not debouch into their opponents' territory either, but by the end of that time the Russians had suffered at least a quarter of a million casualties over the whole of the Eastern Front, and had lost massive quantities of irreplaceable materiel. So that when their capture of Przemysl did at last take place, and they started to push forward towards the Carpathians, they had no immediate resources to withstand the inevitable Austro-German counter-offensive. Przemysl was retaken on 2 June, 1915, and the Russians were driven out of Poland.

Even more significant was the defence of Verdun on the French frontier. A total of 60 forts and *ouvrages* ('works') had been built – and continuously updated –

Austro-Hungarian troops after their recapture of Przemysl, June 1915. The traversing ring of one of the displaced cupolas is clearly visible, but apart from that one mounting, it is again the concrete which has suffered rather than the steel.

between 1874 and 1914. This was far more than had ever been done for the Belgian forts. Moreover, the Verdun forts were better sited, affording each other mutual artillery protection. They were also (as the dates indicate) of more recent – and hence stronger and more resilient – construction. In Belgium, for example, the early users of concrete were ignorant of its properties: they had often poured it in a layer at a time, allowing it to set between each pouring, thus forming internal lines of weakness liable to fracture under stress. Nor had early concrete been reinforced with steel bars or any other form of internal bonding. In addition, the Liège forts had been badly ventilated, and little or no attention had been paid to water supply and sewage disposal.

In contrast, the Verdun forts had been provided with two thicknesses of better, reinforced, concrete with a layer of sand between (making a total of 2.6 m or 8½ ft), the whole covered with earth to a depth of 5.5 m (18 ft). Their cupolas were of 30-cm (11.8-in) steel – not iron; many were retractable, and all mounted guns that were either of larger calibre or of improved performance.

EBEN-EMAEL

THIS BELGIAN fortress was built in the 1930s overlooking the Albert Canal, itself a combined anti-tank obstacle and commercial waterway stretching from Liège 25 km (15 miles) southwards to Antwerp. The fort had its own anti-tank earth-retaining wall, forming a roughly rhomboid trace. Within were 16 75-mm (2.95-in), five 60-mm (2.4-in) and two 120-mm (4.7-in) guns. Some were in casemates, others as single or twin mountings in cupolas that were revolving or retractable or both. There were also armoured positions for grenade throwers, machine-guns and searchlights, plus all the ancilliary equipment, accommodation and stores of any underground fortress.

At 5.20 am on 10 May, 1940 the arrival of nine gliders on Eben-Emael indicated Germany's declaration of war on Belgium. The troops they disgorged used hollow charges to blow in the cupolas and embrasures, smothering others with flamethrowers. In spite of casualties, the garrison improvised barricades in the tunnels and continued firing, not only at the airborne troops on the fort itself but also at the forces trying to cross the Albert Canal bridges. After the arrival of German reinforcements the next morning, at about midday on 11 May, 1940, Eben-Emael surrendered. The 700 Belgians had suffered 82 dead and wounded; the glider-borne Germans 21 casualties out of 55.

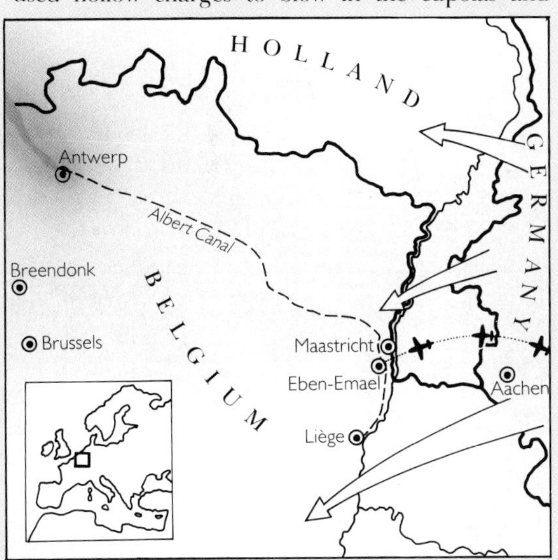

The fortress of Eben-Emael dominated the bridges over the Albert Canal, whose possession was vital for the 1940 German advance into Belgium.

Work begins on reinstating Eben-Emael and the canal bank after its capture in May 1940.

THE ATLANTIC WALL

Festung Europa's ultimate defence against the combined might of the Allied forces

When preparing *Festung Europa* against amphibious assault, the Germans put their faith in linear fortifications concentrated around principal ports and likely beaches, although their propaganda implied that the whole coastline from North Cape in Norway to the Spanish frontier was one continuous wall.

The work was undertaken by *Organisation Todt*, an army auxiliary force of *Bautruppen* (or construction troops). At first it was composed of German volunteers and requisitioned building companies, but later on the bulk of the work was performed by virtual slave labour. Even so, *Organisation Todt* continued to attract specialists from all over occupied Europe. They volunteered either because they admired the Nazis or because the pay was good, or because they were Allied agents, reporting what they saw in otherwise prohibited areas.

An architectural feature of the Atlantic Wall was the massiveness of its structures. Concrete walls and roofs might be up to 3.5 m (11½ ft) thick, their external dimensions magnified by the use of concrete shuttering. Normally, parallel wooden panels are positioned, between which the concrete is poured, and the panels are then removed after the concrete has set. But that takes time, time in which British and American air raids could dislodge the shuttering at a critical stage. So fairly thin pre-cast cement blocks were stacked around the steel reinforcing rods, and the concrete poured in. The shuttering blocks were left *in situ*, adding further solidity and a cushioning effect to the main structure. Curved surfaces had to be sacrificed to the demand for hasty construction, and were replaced by flat faces.

Some German installations were given concrete raft foundations much greater in area than the structures they supported. This was because aerial bombs, missing the precise target and exploding close by below the level of normal foundations, might overturn the structures

Above: North-west Europe was not in fact walled in behind one continuous fortification as many believed. Port approaches were well fortified, especially in Norway, the Pas de Calais and the Channel Islands. Elsewhere the Germans intended to rely mainly on coastal craft, mines, shallow-water obstacles and pillboxes to wreck assault vessels. In the event, the Allies took their own prefabricated port with them.

Below: The advantages of siting pillboxes and other fortifications on a concrete raft.

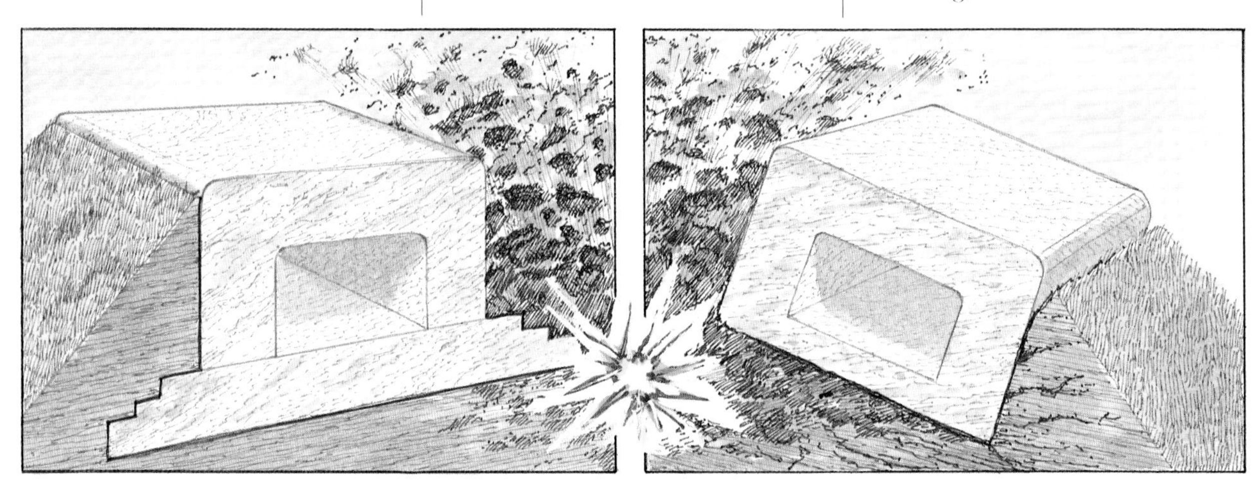

or at least unseat the gun or other equipment within. The 'concrete raft' idea either detonated near-misses above foundation-level or – if lower than that – distributed the shock of explosion evenly throughout the whole structure.

All the casemates, pillboxes and observation towers were fronted by a variety of beach devices for ripping apart landing craft. Their flanks were guarded by minefields, by barbed wire, and by infantry in trenches, foxholes and bunkers.

Among the factors which enabled the Atlantic Wall to be breached when D-Day arrived was the accuracy of heavy naval gunfire, and the employment of specialist tanks. The latter cleared paths through minefields, laid bridges across ditches, planted explosives against obstacles and pillboxes, and spewed flame at anything that looked suspicious.

(The petroleum gel employed by flamethrowers in the European theatre of operations was a thin mixture producing a big bushy flame, its very appearance often persuading defenders to retreat. The Japanese in the Pacific were not so inclined to retreat or surrender, so a thicker gel under greater pressure was employed there. This enabled a smaller-diameter jet of flame to be hosed in through the narrow slits of a bunker, ricocheting round very corner of the interior, igniting everything and everybody within.)

Whether the Germans manning the Atlantic Wall defences in the Channel Islands would have put up such suicidal resistance is unlikely. However, they definitely expected the British to attempt an assault on them, – that is what customarily happens when 'the sacred soil of the homeland' is occupied. As a result, the coastal defence structures on the Channel Islands were the most formidable sections of the Atlantic Wall. They were also the most unscarred by battle. The British declined the challenge, left the garrison alone, and waited until the end of the war to reclaim their own without a fight.

Above: Although captioned by the German Information Service as 'The Atlantic Wall', these troops may be exercising in the Sudetenland fortifications built by Czechoslovakia, but occupied by the Germans in 1938.
Below: *Festung Europa* (Fortress Europe) symbolized: a 40.6-cm (16-inch) gun of Batterie Lindemann in the Pas de Calais, manned by the 244th Naval Coastal Battalion.

MAUNSELL FORTS

FROM 1940-1944 British coastal convoys, particularly in the Thames estuary, suffered badly from attacks by German E-boats and aircraft, and from underwater mines dropped from single aeroplanes at night. Consequently, seven anti-aircraft forts were built in the Thames estuary, and three more off the Mersey estuary. The forts were manned either by the army or by the navy.

The Royal Navy was used to living in ships in which the guns were located on top of the magazine, engine-room and accommodation, with control position, radar and radio on a superstructure between the armament. So the navy -forts comprised all this mounted on top of and inside two cylindrical concrete pillars, themselves supported on one raft. The whole structure was towed out into the estuary and sunk in position, all ready for action.

The army was used to having more space and a wider dispersal of facilities, so army forts each comprised seven octagonal two-storeyed houses, each one mounted on four concrete legs rising from an open-framed concrete pontoon. All seven were towed out separately and sunk in position about 40 m (44 yd) apart. They were then linked by open catwalks some 15 m (50 ft) above the sea.

Both army and navy forts were designed by the civil engineer G. A. Maunsell, and both types mounted 94-mm (3.7-in) and 40-mm (1½ in) anti-aircraft guns.

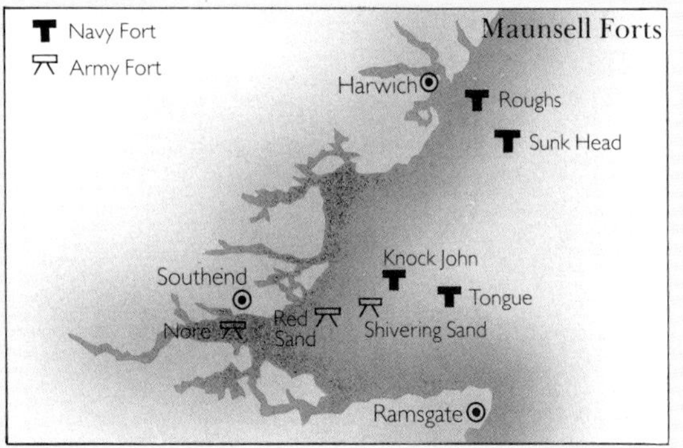

Above: A War Department supply tender alongside the radar-controlled searchlight tower of one of the British Army's Maunsell forts in the Thames Estuary. Moving right to left the other towers are: 40-mm Bofors (1.58-inch) mounting; 3.7-inch (94-mm) mounting; central control tower (the wartime censor has whitened out the latest electronic aerials); and three more 3.7-inch (94-mm) mountings. 150 officers and other ranks could be accommodated in this fort, providing round-the-clock and all-weather anti-aircraft defence in their area. Army and Navy forts together accounted for 22 aircraft, over 20 flying bombs, and an E-boat, plus rescuing ditched Allied airmen.

The Verdun forts were even proof against *Big Bertha*, as the garrison of Fort Douaumont learned early in 1915; it was not an experience they particularly cared for, but they had survived – and so had Fort Douaumont. Nevertheless, the French Commander-in-Chief, Marshal Joffre, had not been impressed with the Belgian forts. And he was disillusioned with the whole idea of locking up thousands of troops in what were obvious targets and all too often death-traps. He preferred his soldiers to be out in the open, or at least in less obvious trenches. There they would form a manoeuvrable (if not mobile) field army, all ready to march into Germany – he did not want anyone getting a siege- or defence-mentality. So Joffre ordered the Verdun forts to be dismantled and the troops transferred.

A demolition team was actually in Fort Douaumont when the Germans launched their massive offensive on the Verdun Front on 21 February, 1916. One of their first successes was the capture of the otherwise empty Fort Douaumont. But the other forts proved more difficult to overwhelm. Fort Vaux, just 3 km (2 miles) away, did not fall until 6-7 June, 1916. A month later the German offensive reached its limit. By 18 December, 1916 the two lost forts had been recaptured; Verdun had been held; and the Germans (suffering almost half a million casualties) had not passed further into France. But it had cost France more than half a million men dead, wounded and missing. It was the unceasing replacement of men in the supporting

Above: Looking eastwards into Germany and muffled against the bitter cold of January/February 1940, French infantry have used gabions to improvise an outpost line forward of the Maginot fortifications.

Below: Anticipating Allied invasion, the Germans occupying Le Havre built this steel-reinforced concrete pillbox into an existing harbourside hotel. After liberation on 12 September 1944, the surrounding house has been destroyed by aerial and naval bombardment, but the paint-disguised strongpoint still survives – as does the air-raid shelter in the background.

trenches as fast as they were taken out of action that saved Verdun, not the actual fortresses themselves.

Nevertheless, those forts had been solid strongpoints, providing secure fire-bases and infantry shelters from which aid could be speedily delivered to temporary fieldworks, which in turn prevented small teams of assault engineers from getting within striking distance of fortress cupolas, casemates and entrances.

In fact, all the fieldworks of the Western Front created two confronting linear fortresses which could not be stormed by frontal assault except at the cost of prohibitive casualties. If this could be achieved by the hurried excavation and erection of temporary obstacles, how much more effective their replication in permanent materials ought to be.

That was the reasoning behind the Germans' Hindenburg Line. They did not have time to build fortresses of pre-1914 dimensions, but between 16 September, 1916 and 23 February 1917, they created what Hindenburg called *Die Siegfriedstellung*. Its northern terminus was on the existing front line near Vimy Ridge and Arras, its southern end beyond St Quentin. It comprised a series of trenches running parallel to the proposed new front line, which incorporated all the stratagems of trench warfare and were soundly constructed with timber, sandbag or concrete revetment.

Left: World War II strongpoints and fieldworks were established everywhere, their design varying according to the surrounding terrain and local materials. This double pillbox was built by the Germans in Yugoslavia, resembling the hilltop towers of an earlier period.

Above: A 1940 British pillbox near Dover – a Type 26 variant, accommodating five men and three LMGs, such as the .303-inch (7.7-mm) Bren. Erosion has exposed its raft foundation and basement lower structure needed for level siting on a slope.

Left: France, late summer, 1944. Tanks and infantry could not pass these concrete-filled pilings and barbed-wire entanglements, especially as in action they would be fitted with booby-traps and covered by the blockhouse at the base of the reinforced, firing-slitted tower, and by the pillbox built into the corner shops opposite. Soon these German prisoners will be clearing their own devices.

Below: 1943 practice shoot by battleship *Rodney's* 16-inch (406-mm) guns, very effective against coastal fortifications.

Right: Flat-bottomed railway track and other lengths of old steel form anti-tank obstacles in the Maginot Line. Note the observation cupola over the distant casemate. (It can be assumed that the area is not mined, otherwise the photographer would not be standing there.)

Below: The 39th Infantry Regiment (US 9th Division – helmet badges censored) ride a Sherman bulldozer through the interconnected dragon's teeth of the Scharnhorst (or Siegfried) Line north of Roetgen near Aachen in September 1944.

The front, support and reserve trenches were connected by communication trenches and angled switchlines that could be used as further defensive positions if the British or French managed to occupy one sector of the front trench. At regular intervals further communications networks led back to underground complexes containing headquarters, signals units, accommodation blocks, kitchens, hospitals, power-stations, stores and magazines. Each structure's reinforced-concrete roof could have a thickness of 1 m (3¼ ft) and be as much as 7.5 m (25 ft) below the battlefield. All were fitted with hidden mines and boobytraps which, when armed by the last man in retreat, could then be detonated by timer or trip-wire.

Scattered about on the surface – apparently in random fashion, but all with overlapping, mutually-covering arcs of fire – were pillboxes mounting machine-guns and light artillery. (These strongpoints became features of all battlefields and potential battlefields during World War II.)

The Allied trench systems of 1915-17 were similar, but nowhere near as complex nor as permanent. It was considered morally unacceptable to encourage defence – it was too close to defeatism: Allied troops had to be offensively-minded. And there was good reason for this: all the Germans had to do to win World War I was to sit where they were. On the other hand, French and Belgian victory could be achieved only by

Below (top): Post-World War II 5.25-inch (133-mm) anti-aircraft gun emplacement, North Front, Gibraltar.

Below (bottom): Though knocked out by the British cruisers *Ajax* and *Argonaut* at 0845 on 6 June 1944, the 15.2-cm (6-inch) battery at Longues still overlooks Gold Beach in Normandy.

Right: Chouet Tower on Guernsey, one of a chain of naval direction-finding and observation towers, which in wartime were topped with aerials to tell the Germans about shipping movements in the Channel. There is an Army coastal artillery observation post on the right.

expelling the invader, and the British forces were in Europe to help them do just that.

Unlike the Allies, therefore, who were not prepared to abandon any more of their homelands, the Germans were willing to retreat if that made it easier for them to hold the rest more securely. And that is what they did in February-March, 1917, laying waste the countryside as they did so. The Hindenburg Line was not totally impregnable – it was penetrated on a number of occasions before the final victory – but it did make for a more economical use of manpower while also inflicting costly casualties on a tiring enemy.

If this could be done in the urgency of wartime, actually on an active battlefront, how much more effective it would be if such work were designed and implemented in peacetime, or at least in a comparatively quiet area before the real war arrived.

This was the reasoning behind the Sudetenland fortifications, which so effectively barred the German route into Czechoslovakia that Hitler was forced to

provoke the Munich Crisis of 1938 – a gambler's bluff that won him that particular prize.

This was, too, the reasoning behind the Maginot Line, which so effectively barred the way into France in 1940 that the Germans were forced to outflank it, risking battle with the field armies of Holland, Belgium, Britain and France – another gamble that succeeded.

And this was also the reasoning behind the Atlantic Wall, which was intended to stop the Allied assault on the invasion beaches of north-west Europe. This time the fortification was not effective. By 1944 it was not enough merely to strengthen bunkers, gun-sites, observation posts, machine-gun nests and soldiers' shelters against horizontal or near-horizontal fire; all these had also to be protected against bombs dropped from the air.

By 1944 *Festung Europa*, like every industrialized society, had to have an 'invisible roof' to protect *all* its factories, its communications and its natural resources (including its human resources) against aerial bombardment and airborne assault. There had to be shelters to protect those civilians upon whom war manufacture depended. There had to be anti-aircraft gun emplacements, airfields, hangars, control towers, radar installations and wireless masts, all the weaponry of three-dimensional and electronic warfare, and all the equipment and fortifications of twentieth-century civil defence. The system had to span the whole continent and had to be completely impenetrable to be effective.

Left: The ruins of 'Caesar' (No.3 Casemate of Batterie Lindemann at Sangatte in the Pas de Calais. Half-ton (600-kilogram) shells fired from here could reach 34 miles (55 kilometres) across the Channel.

'Caesar' under construction. Note shuttering, camouflage and muzzle-cover (keeping salt damp out) easily removed by the lanyard visible right.

Chapter Six

THE SHRINKING WORLD

For a while, after 1945, it seemed as though there was no place for fortification in modern warfare. Certainly, traditional coastal defence came to an end. There was no point in mounting long-range artillery to fire at battleships 40 km (25 miles) away – not when major powers had nuclear-armed aircraft and missiles which could sink those same battleships as soon as they left harbour, or even while still in harbour, on the other side of the ocean.

Yet the threat of cataclysmic war did not mean the end of war itself. Hostilities have continued in various theatres almost without pause since 1945. Whether called police action, counter-insurgency, freedom struggle, the restoration of law and order, keeping the peace, the provision of military aid, terrorism, or the re-establishment of national sovereignty, they have all been conflicts in which every type of weaponry has been employed, including improvised and permanent fortifications – and some conflicts have seen the re-use of very ancient fortifications.

The Navy controls Sweden's coastal defence. This 1970s turreted 75-mm (3-inch) gun has a range of 7½ miles (12 kilometres). Most emplacements are camouflaged, with the commander in an armoured cupola.

THE MIDDLE EAST

Officially entrusted to Great Britain under League of Nations mandate following World War I, Palestine became effectively part of the British Empire. The departure of the occupying British forces in 1948 was the signal for open conflict to begin on 15 May. The newly established state of Israel was assailed by Egypt, Jordan and Syria, themselves coming to the aid of the threatened Palestinians of both the Muslim and Christian religions. Both Jews and Arabs fought for control of the Old City of Jerusalem, which contained the site of the Temple, the Dome of the Rock and the El Aksa Mosque. While engaged in this operation, the Arab Legion was simultaneously having to use the cover provided by the crenellations of the ancient walls of the Old City to fight off the Jewish besiegers holding the New City. The latter were themselves cut off by Arab and Egyptian field armies. The Arab Legion gained

Left: German technician preparing a V2 ballistic missile for launch from its mobile platform. These single-stage, liquid-fuel rockets entered service in 1944, 1,371 being directed at London and Antwerp.

Below: Battle of Britain radar installation at Poling, Sussex. The transmitting aerials were slung between the metal pylons (left); the receiving aerials were fixed directly on the wooden towers (right).

control of the Old City on 28 May, 1948 but were unable to make headway against Israeli positions in the New City. It was not until 11 June, 1948 that the latter were reached by supplies and reinforcements which had fought their way along the only road from Tel Aviv (although some items had been dropped from improvised military transport aircraft).

The battle within the city was mainly one of infantry weapons, partly because the forces engaged were short of heavier equipment, partly because even if it had been available it was not suitable for house-to-house and street-fighting, and partly because both sides wanted to occupy Jerusalem, not to destroy it. In the subsequent armistice agreement, Jerusalem was divided between Israel and Jordan.

The Six-Day War began on 5 June, 1967. A series of Jordanian fieldworks, bunkers and barbed-wire were overcome by Israeli infantry armed with grenades and automatic weapons, and supported by artillery, tanks and reconnaissance vehicles. During the night the Israeli assault pushed through the suburbs, but stiff opposition prevented them from reaching the Old City until dawn on 6 June. The Old Wall still proved its defensive strength against infantry weapons, the Arab Legion making skilful use of merlon and crenel. Recoilless artillery and tanks were brought up from the northeast and through the outer city streets; guns and automatic weapons swept the wallwalk and hammered at St Stephen's Gate – taking care not to hit any of the Holy Places. Early the next morning an Israeli half-track command vehicle pushed its way through St Stephen's Gate and infantry began eliminating the last Arab army snipers holding out in the Old City of Jerusalem.

Meanwhile, other Israeli forces had advanced to the River Jordan in the east and to the Suez Canal in the south-west. Quite apart from the political significance of the acquisition of this territory, Israel's military aim was to secure defensible borders.

Below: The debris of modern battle: wrecked Egyptian tanks in the desert after the Six Day War of 1967.
Right: The fruits of victory. Israel's Chief of Staff Yitzhak Rabin, General Moshe Dayan and General Uzi Narkiss walk through the ancient fortifications of Jerusalem.

There have been subsequent adjustments (a feeble term for lengthy periods of open hostilities and delicate diplomatic negotiations), the most significant being the return of the Sinai Peninsula to Egypt, but at present Israel's frontiers run along the Sinai Desert, the River Jordan and the northern mountains. All these are comparatively easily defensible, especially when reinforced with a 'defensive wall' of fieldworks continuously maintained and updated with prefabricated components, land-mines and surveillance equipment, backed up by mobile forces on land, at sea and in the air. Israel is too small an area to permit defence in depth – except at the expense of neighbours, an internationally unacceptable policy – so the underlying strategy has been to turn the whole state into a 'walled fortress'.

Walls can be breached, however, as happened to Israel's Bar-Lev Line under Egyptian assault in 1973.

Linear defences are not necessarily economical in manpower, either, as the French found in 1957. To try to prevent insurgent infiltration into Algeria from Morocco and Tunisia, they erected a series of massed barbed-wire entanglements and deeply buried anti-personnel mines from the coast inland for 300 km (200 miles). Parallel military roads enabled army units to head off the survivors of attempts to get through these obstacles; their task was eased by electric fences, any interference with which automatically triggered an alarm for that section. Even so, these defences required the deployment of some 150,000 troops just to garrison and patrol these barriers. Nevertheless, the campaign proved a success, especially when combined with air patrols over the desert beyond the ends of the Morice Lines. However, *le Front de Libération Nationale* (the FLN) maintained the political advantage, and in the end Algeria became independent of France.

Defensive lines were also established in the Dhofar coastal region of southern Arabia in 1973. Again combined with air patrols beyond the ends of the 30-km (50-mile) Hornbeam Line and the smaller Damavand Line, they completely blocked the movement of camel columns carrying munitions for rebels from the Yemen into Oman.

The most extensive of these desert counter-insurgency linear-barrier systems was the 3,000 km (2,000 miles) of dry-stone walls, 3 m (10 ft) thick and 2.7 m (9 ft) high, stretching across Western Sahara. They were built by Moroccan troops in the 1980s to prevent the rapid movement of Polisario guerrillas operating from Algeria. Although the infiltrators were well equipped with 36.5-tonne T55 tanks and armoured personnel carriers, even these tracked machines experienced difficulties in scrambling over or butting through piles of jagged rocks; for ordinary wheeled vehicles it was totally impossible. Radar and ground sensors even picked up approaching infantry. Any Polisario who did get through were trapped against the next wall by the mobile Moroccan forces operating in that zone (similar to the 'drives' towards lines of barbed-wire and blockhouses conducted by the British during the Boer War). Polisario too distant from a patrol to be overtaken were located and strafed by Moroccan aircraft.

Such stone walls constitute an excellent example of how even simple linear fortifications have been employed effectively throughout history to discourage mobile enemies. All that has been required is the labour

A Vietnamese soldier raises the flag over Dien Bien Phu at 1730, 7 May 1954. But even at the moment of victory, his two comrades still keep watch. Note the PSP (pierced steel planking), providing roads and runways on poor terrain and the corrugated steel roofing over the entrenchments, themselves revetted with posts, boards and stones, and protected with barbed wire. Note too the surrounding hills, from which Vietnamese artillery poured shellfire down upon the French fieldworks and emplacements.

and resources to build them in the first place, and the defenders' ability to hurry to an advantageous battle-field of their choosing.

VIETNAM

Where the terrain does not favour mobility either by a defender or an attacker, however, a better strategy might be to fortify some feature that controls the only communications in the area. That is what the French did at Dien Bien Phu in Indo-China. General Castries established some 14,000 troops in a valley ringed with fieldwork strongpoints, all with overlapping fields of fire. Their supplies of ammunition were guaranteed by two airstrips, ensuring that deliveries could not be interrupted because the guerrillas had no air force.

However, Dien Bien Phu was ringed by hills. The Viet-Minh may not have had aircraft, but they did have 200 field-guns and mortars. And the terrain was not impassable to General Giap's 72,000 fighting men. On 13 March, 1954 the Viet-Minh artillery looking down on the French emplacements opened fire. One after the other, the strongpoints were knocked out or overrun, as were the two airstrips. Supply-drops by French planes could not replenish fast-disappearing ammunition as the French fell back within a shrinking perimeter. Their last position was overwhelmed on 7 May, 1954: 2,293 Frenchmen had been killed; the rest (including 5,000 wounded) marched into captivity.

The importance of high ground in this sort of situation was not lost on the Americans when they fought their war in Vietnam in the 1960s. In lowland

A rear-loading Boeing Vertol CH-47 Chinook brings in supplies and equipment (especially barbed wire) for Americans establishing a hilltop outpost during their 1960s war in Vietnam. The dust will turn to mud when it rains.

areas, artillery units (usually self-propelled) had of course to be emplaced on level ground. But if at all possible, a hilltop site was selected for a fire-support base. Sometimes a site was so inaccessible that everything, including 155-mm (6.1-in) guns weighing up to ten tons, had to be lifted in by helicopter. The heaviest loads were carried by Sikorsky CH-54 Skycrane. The pieces of heavy equipment most commonly lifted in (delivered by Boeing-Vertol CH-47 Chinooks) were M-102 howitzers with a calibre of 105 mm (4.1 in) a range of 11½ km (7¼ miles). The guns would be brought in ready for action, sometimes even while the position was still being secured.

As many as six howitzers would be emplaced, each enclosed within its own circular revetment and jacked up in such a manner that it had all-round traverse. Each gun-crew had its own log-built, earth-covered, sandbagged shelter, as did the command and communications positions close by. The observation tower was inevitably more exposed, and accordingly built as strongly as it possibly could be. There were also simple catering and first-aid facilities. Around the base were sited 16 infantry positions, looking like sandbagged pillboxes. These were equipped with recoilless rifles, machine-guns and grenade launchers. Two of these positions flanked the only gate through the barbed-wire stockade.

Hidden not only within the entanglement but also outside *and* just inside, were trip-wires and anti-personnel mines. All the surrounding areas had been cut down, and growth farther out was sprayed with defoliant chemicals. (Whatever the region, every fire-support base in Vietnam was thick with dust in the dry season and deep in mud in the wet.)

In spite of these precautions, fire-support bases (FSBs) were frequent targets for Viet-Cong attack. Hand-held rocket launchers caused havoc in the tightly-packed strongholds, especially during helicopter resupply operations. In fact, the FSBs could not have existed at all without total air superiority – nor if they had been subjected to electronic countermeasures.

Everything depended on radio communications. An American infantry patrol many miles away in the

Below: 'A Tunnel Rat' (Sergeant Ronald Payne) during Operation Cedar Falls, 25 miles north of Saigon, January 1967.

Left: Early 1970: in heavy rain a Chinook air-supplies the 4th Battalion of the 31st Infantry Regiment (196 Brigade) at their base established by the American Division west of Tam Ky. Note the internal barbed-wire entanglements to prevent any infiltrating Viet-Cong from rampaging through the camp.

Above: Mobile warfare in the nuclear age: Vietnamese tanks and infantry push on to Saigon (renamed Ho Chi Minh City in 1975).

jungle might find itself under sniper fire. Messages to the FSB brought single ranging shots nearer the target until the whole battery crashed down upon the enemy. Similar techniques were used by the fire-support bases themselves to call down shells from neighbouring FSBs upon Viet-Cong assailants too close to the perimeter for their own weapons to bear.

All this high-technology weaponry contrasted strangely with the underground fortresses built by the Viet-Cong. Wherever the subsoil permitted, village cadres using hoe-blades, coca-cola tins and their hands excavated complex tunnel systems, dumping the earth far away from the workings. There were sleeping-quarters, dumps of food and ammunition – grandiose terms for perhaps some sacking, half a bag of rotting rice and a clip of rifle cartridges – yet men, and women, existed down there. Using such equipment – and what could be stolen from the Americans or scavenged from abandoned camps – the Viet-Cong emerged again and again to carry out rocket attacks and sniper ambushes in areas that had been apparently cleared many times of the enemy.

The only way to deal with these tunnels, and their occupants, was for small teams of small-statured Americans to crawl one at a time into each near-vertical hole in the ground. Armed only with pistol, knife and flashlight, they never knew when their hands and knees might press down on excreta-poisoned *punji* stakes, when a shoulder might snag the wire pulling the pin out of a hand-grenade, or when a movement of the head might dislodge the carefully-positioned pot containing venomous snakes or insects. Sometimes the tunnel would come to a dead-end, so that the soldiers had to crawl backwards, their retreat now cut off by a hidden trap-door from which emerged the enemy.

Below: The deployment of missile bases, where the US Cavalry once rode out from frontier forts. (The Minuteman was named after the patriots of the War of Independence, ready to turn out at a minute's notice.)

Right: North American Air Defense (NORAD) Headquarters under the Cheyenne Mountains in Montana. The coordinates on the map are of latitude and longitude, and do not mark the actual positions of Distant Early Warning sites.

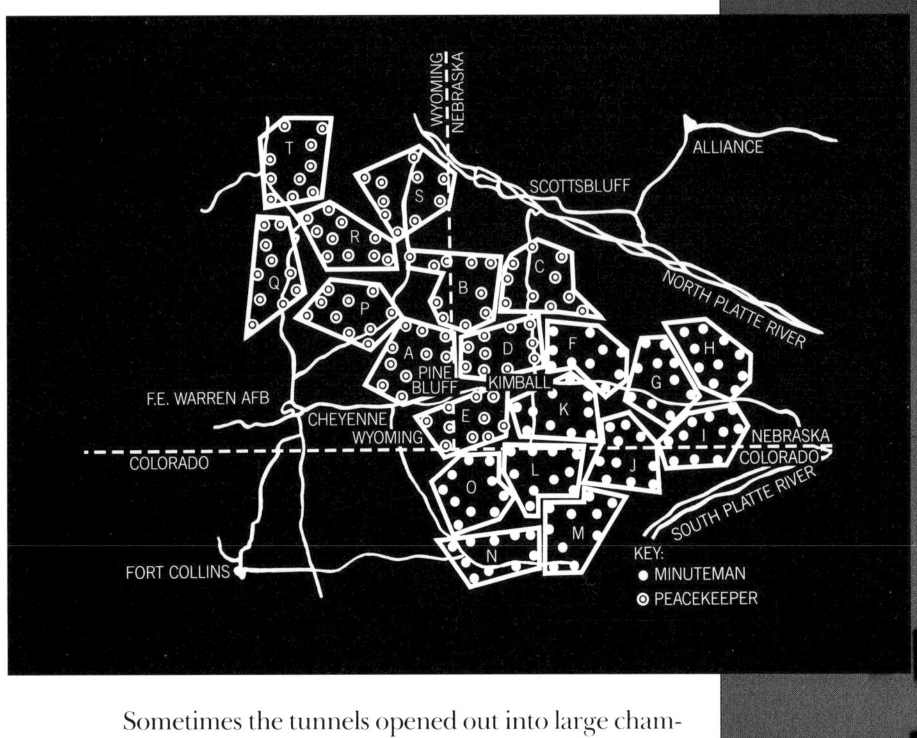

Sometimes the tunnels opened out into large chambers with a well and cooking stove (the smoke led far away before reaching the open air). There might be underground latrines or first-aid posts. One tunnel complex had an operating theatre, its earth ceiling and walls draped with disinfectant-impregnated (ex-American) parachutes. Sometimes the tunnels curved round so that the 'tunnel rat' found himself exiting right behind his comrades clustered round the hole where he had gone in.

The outcome was invariably the same: explosives were planted or gas was sprayed in – and another 100 m (100 yd) of tunnel had been destroyed . . . out of a total of 240 km (150 miles) of Viet-Cong tunnels. Some of these complexes were even under the biggest American camps in Vietnam.

Primitive as they were, such tunnels fulfilled the function of strongholds throughout the ages: they provided refuge from attack and served as bases for offensive operations.

FORTIFICATION FOR THE FINAL CONFLICT

In all these conflicts (and all others since 1945) fortification has played some part, even though it might have been in the form of prefabricated yet temporary fieldworks.

GMT
LOCAL
NRINT

Saab J-35 Draken jet fighters of the Royal Swedish Air Force taxi out from their underground hangar.

In none of these conflicts have nuclear devices been employed. Instead, the nuclear weapons have remained sinister but dormant in their own forms of fortification. Missiles are housed in underground silos, vertical tubes 25-50 m (80-165 ft) deep, surrounded by the necessary apparatus for launch. All that is visible above ground is an upturned saucer of concrete 8 m (25 ft) thick. There is a small trap-door in this carapace for personnel to gain access for occasional inspection of the silo and its missile, but normally the whole underground complex is left unmanned, the missile always fuelled and ready for flight.

The flight launch control centre (also underground and built to resist the shock of subsoil explosions) may be anything up to 5½ km (3½ miles) distant. Upon receipt of an order to launch – and after the various fail-

Draped with camouflage netting so that it looks just like another dark rock on the Swedish hillside, a 75-mm (3-inch) coastal defence gun fires out to sea.

safe procedures have established and confirmed full authorization – the two officers on watch activate the launch sequence of the ballistic missile under their command. A segment of the circular carapace is robotically pulled back, and within a minute 'the bird' is climbing away at 24,000 kph (15,000 mph) on its irreversible journey up into space and down on to its target 12,000 km (7,500 miles) distant.

These nuclear-warheaded intercontinental ballistic missiles have marked out the global chessboard on which conventional wars can be fought. Combatants in such wars may be professional task forces or guerrilla units, massed conscript armies or even garrison troops in traditional fortifications – but they must conduct their operations according to certain rules and with certain limited objectives that can be obtained by

Above: Swedish submarine *Nacken* leaving her underground dock.
Left: 8 February 1982: LCC and TEL vehicles (Launch Control Center and Transporter Erector Launcher) on Californian mountain roads near San Diego – a practice deployment of the General Dynamics Tomahawk Cruise Missile with a contour-hugging range of 2,000 miles (3,000 kilometres) at 550 miles (880 kilometres) per hour. Although this mobility is the very antithesis of static fortification, such missiles have prompted the construction of nuclear shelters. When not on the road, they themselves require military architecture for storage and maintenance.
Right: Artist's impression of the SS-20, a Russian missile, on a mobile launcher.

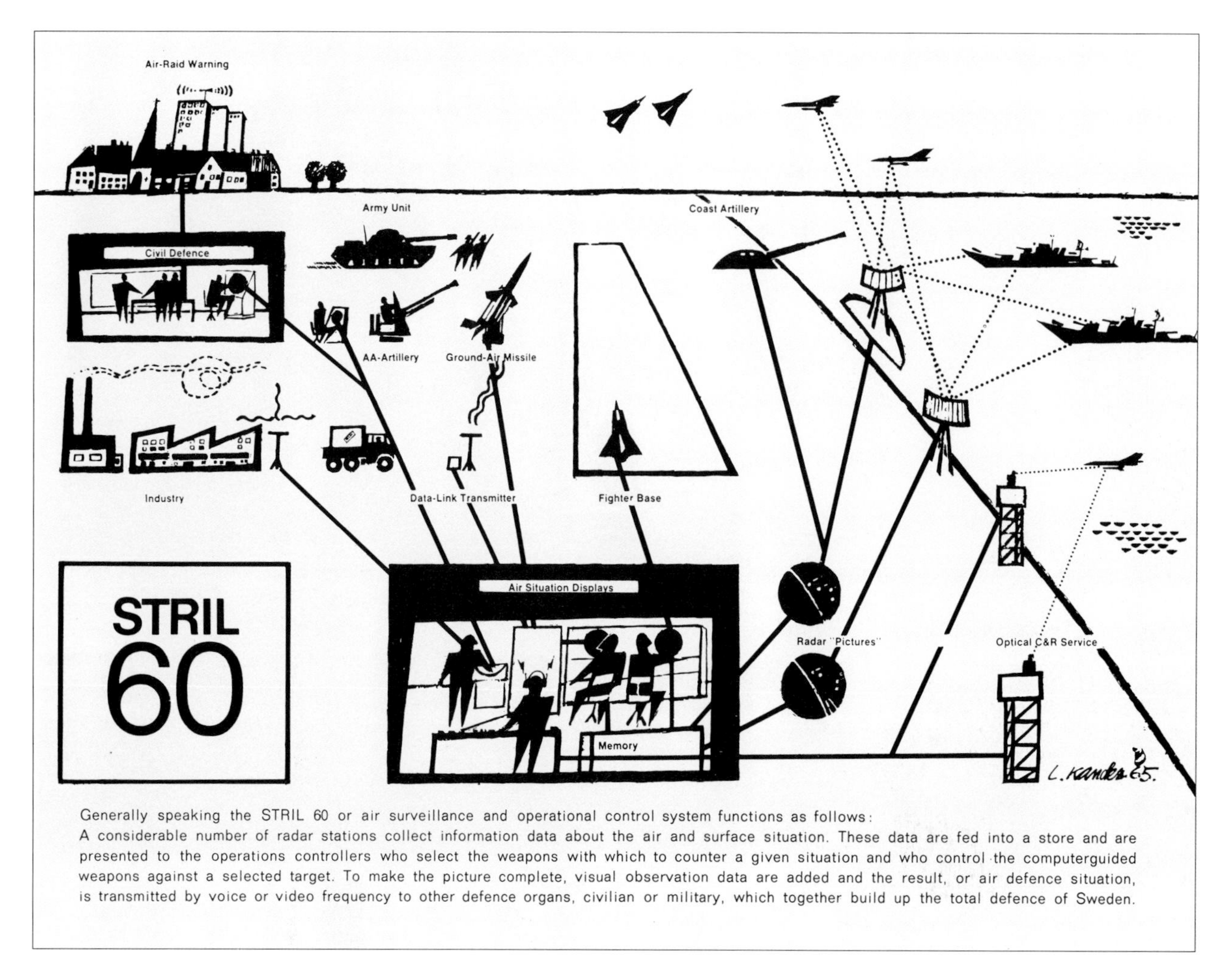

Generally speaking the STRIL 60 or air surveillance and operational control system functions as follows: A considerable number of radar stations collect information data about the air and surface situation. These data are fed into a store and are presented to the operations controllers who select the weapons with which to counter a given situation and who control the computerguided weapons against a selected target. To make the picture complete, visual observation data are added and the result, or air defence situation, is transmitted by voice or video frequency to other defence organs, civilian or military, which together build up the total defence of Sweden.

diplomatic or subversive means if these subsequently prove more expedient. The people directing each war know that if they break the rules and allow the conflict to escalate, they risk the ultimate punishment of irreversible global holocaust, in which they too will die along with their subjects and fighting forces.

In case that does happen, most of the world's leaders have prepared underground shelters for themselves and for the handful of soldiers, scientists and administrators who will eventually govern what is left outside. The shelters are impregnable hideouts, equipped with the means of producing their own air, some even with the means for the inhabitants to live off their own recycled waste products. All the armoured filter-ventilators, observation-slits and remote-control cameras necessary for monitoring external conditions are proofed against radioactive, chemical and bacteriological assault. Provisioned with food and drink for months, most shelters also have facilities for entertainment and sporting exercise. Sealed-off sections may contain works of art, the Crown Jewels and bars of gold, preserving for posterity – no matter how few – the priceless treasures of civilization as we know it . . . provided the designers have remembered to build the toilets *inside*, and have used a roofing material that stops the rain getting in.

Private citizens will either be abandoned to nuclear destruction or be provided with some form of personal shelter, both provision and quality depending on their government's resources and attitude.

But all that is hypothetical, literally futuristic, like the Strategic Defence Initiative, in which electronic equipment erects invisible 'walls' around whole continents, automatically triggering missile-response to intercept and destroy incoming rockets. Of more everyday relevance are the frontier barriers like the Berlin Wall, which was erected as much to keep law-abiding citizens in as to keep an enemy out.

It is ironic that at a time when such international obstacles between Communist countries and the rest of the world seem to be diminishing in significance, all government agencies – even military organizations – are having to fortify their buildings against invasion by their own disgruntled subjects. External walls are fringed with barbed wire; windows are made of armoured glass or completely blocked off by steel shutters; unseen eyes inspect the interloper; electronic detectors frisk his person. Turnstile-portcullises only

Far left: STRIL-60, the surveillance and operations system of the Royal Swedish Air Force. Information from electronic and visual stations is fed into the central control, where it is evaluated and appropriate action taken – to scramble aircraft, order coastal artillery to open fire, send civilians to fallout shelters...or decide that the approaching ship or aircraft is friendly. The latter can be done by one electronic device automatically interrogating the suspect blip on the screen – which *has* to be done (and counter-weapons launched) within seconds, if the intruder is an unmanned missile.
Left: The control room in a high-security organization.
Below: Closed-circuit TV camera – the watching eye.

operate after the computer has accepted the correct plastic-card password. Everywhere there are guards, some discreetly – and some openly – armed.

This atmosphere of apprehension, this anticipation of intrusion, has now spread to private citizens in virtually all countries. Their homes can be protected with all the close-range devices of modern siege-warfare. In fact, many people are in danger of developing a siege-mentality verging on paranoia. It is another curious irony that at a time when humankind can leave behind the confines of this planet, when speedy communications have shrunk the globe into one indivisible world, people are nevertheless turning in upon themselves. Governmental emphasis on personal security tends to foster suspicion of the stranger, contempt for the foreigner, fear of the alien, and resentment of the tourist. The world is shrinking; homes and minds are being barred against the outsider and his ideas.

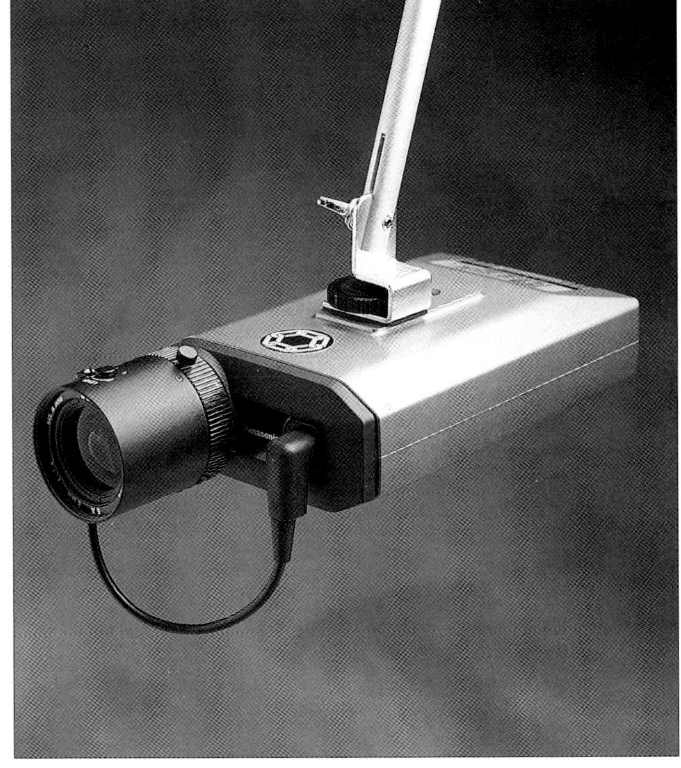

GLOSSARY

Agger A Roman ramp of earth and hurdles, reinforced with tree-trunks, and fireproofed with dry-stone facing. It was designed to reach to the enemy's *ramparts* and walls, enabling siege equipment to be brought into action at close range.

Alcazaba A square Arab fort, built of rough concrete, with square towers at each corner and protecting the gate.

Alcazar A defended palace for an Arab military regional governor.

Allure Walkway along the top of a castle wall.

Arradah An Indian catapult weapon.

Bailey A large clear area, enclosed by a ditch and rampart, or by a moat and wall, and either adjacent to, or surrounding, the *motte* of a Norman castle. Besides providing outer defence, the bailey served as an exercise area, parade ground and an emergency corral for stock. There were also huts for soldiers and skilled artisans.

Ballista (plural: Ballistae) Roman catapult weapon which hurled stones and lumps of timber up and over the enemy walls. Propulsion was provided by a wooden arm embedded in a mass of twisted sinew or hair. The arm was wound down and back, and the projectile placed in a sling or cup at the end. The arm was released, flew forward and hit a massive cross-timber, whereupon one end of the sling flew free and the missile shot through the air.

Barbican A fortified extension to the gateway into a castle.

Bartizan A small tower, usually cylindrical, which springs from the wall or main tower, as opposed to being mounted on top of the wall or tower.

Bastion Part of a fortified wall which projects beyond the line of the wall itself, but is the same height as – or lower than – that wall. If it is higher, then it is a tower.

Batter A sloping stone apron or plinth in front of a castle wall or tower; also called a *talus*.

Blockhouse Small square fortification, usually of timber.

Bonnet A freestanding fortification, providing extra fire in front of a Vauban-period fortress; also called a *priest's cap*.

Brochs Round towers built in northern Britain as refuges from slave-traders and other raiders during the Iron and Dark Ages.

Burg (plural: Burgen) A German stronghold.

Burh A Saxon stronghold.

Caisson A huge cylinder containing high-pressure air. Sunk on to the sea- or river-bed, it keeps the water out. Workers enter via airlocks and can labour in a comparatively dry environment.

Casemates A system of artillery emplacement, in which each piece is individually mounted in its own protected room. It is thus different from a battery, in which cannon are lined up side by side.

Castellum (plural: Castella) The Latin word for a little camp, given to Roman *bastions* and *redoubts*.

Castra A temporary Roman camp.

Castra stativa A semi-permanent Roman camp, similar to *castra*, but strengthened with stone walls, wooden stockades and *castella*.

Charkh A large Indian crossbow.

Chemise Wall Formed by a series of interlinked or overlapping semicircular *bastions*.

Circumvallatio Entrenchment dug by the Romans all the way around an enemy stronghold to prevent the defenders' escape. Usually a simple ditch and bank facing inwards, more complex ones had several entrenchments, *ramparts*, stakes, *palisades* and *redoubts*. The exact opposite of a *contravallatio*.

Cofferdam A wall, usually of timber or metal piling, sunk into the bed of a river or the sea. It forms a watertight enclosure from which the water can be pumped out, thus enabling work to be undertaken in a comparatively dry environment.

Contravallatio An entrenchment dug by the Romans all the way around an enemy stronghold to protect themselves against attack by the enemy's field army. Usually a simple ditch and bank facing outwards, more complex ones had several trenches, *ramparts*, stakes, *palisades* and *redoubts*. The exact opposite of a *circumvallatio*.

Counterguard A long, near-triangular, freestanding fortification, to provide extra fire within the moat of a Vauban-period fortress.

Counterscarp A gallery in the far bank of the moat of a Vauban-period fortress.

Crannog The word used in Celtic Scotland and Ireland for a timber-built, fortified lake village. They date from the Iron Age to the sixteenth century AD.

Crenellation Battlements.

Crenels The gaps in battlements; the opposite of *merlons*.

Below left: Bad Godesberg on the Rhine.
Below: Gravensteen Castle in Ghent, Belgium.

Cross-and-orb The result of Renaissance hand-gunners cutting a round hole (to aim and fire their pieces) at the base of a combination of vertical and horizontal slits in a castle wall, originally used for aiming longbows and crossbows.

Crownwork A freestanding bastioned fortification, providing extra fire in front of a Vauban-period fortress.

Cupola A hemispherical armoured roof.

Curtain Wall The wall joining two or more castle towers.

Cyclopean Masonry Dry-stone walling with blocks so huge and so old that the Ancient Greeks believed that they must have been erected by the mythical giant Cyclops; later applied to all such masonry.

Dead-Ground The area where your bow or gun cannot be aimed, usually because it is too close for you to do so without exposing yourself.

Demi-Lune A freestanding, near-triangular, quadrilateral fortification with a concave rear face, located in front of the arrowhead *bastion* of a Vauban-period fortress; also called a half-moon or lunette.

Donjon The principal fortification of a Norman castle; also known as *keep* or great tower.

Double-Tenaille A small freestanding bastioned fortification providing extra fire in front of a Vauban-period fortress.

Drawbridge A bridge which can be raised or slid back to deny the enemy passage over ditch or *moat*.

Drum-Tower A round tower.

Embrasure A hole cut in a wall or parapet to enable cannon to be traversed and fired.

Enceinte The whole fortress complex.

Fascine A huge bundle of brushwood for revetting ramparts or filling in ditches.

Fire Support Base (FSB) American artillery complexes in Vietnam providing gunfire in support of infantry patrolling the surrounding jungle. Each of six howitzers was emplaced within its own circular *revetment* of sandbags, earth and timber. There were also personnel shelters, observation, command, communications, and infantry defence positions, plus catering and first aid facilities, all of similar construction. The whole area was surrounded by a heavily-mined barbed wire entanglement.

Gabion An earth-filled, cylindrical basket, used for revetting trench walls and gunpits.

Glacis A broad, sloping expanse of naked rock or bare earth on which an approaching enemy is completely exposed to view and to the defenders' weaponry.

Hirojiro A Japanese lowland stronghold.

Hornwork A freestanding, quadrilateral fortification providing extra fire in front of a Vauban-period fortress.

Below left: Château d'If: real-life prison for the fictional Count of Monte Cristo.
Below right: Château of St Malo, in Brittany.

Howdah Passenger-carrying structure on top of an elephant.

Keep The principal fortification of a medieval castle; also known as *donjon* or great tower; sometimes applied to the central fortification of a Tudor castle.

Killing-Ground An area where the enemy is most exposed to the defenders' weaponry with little chance of escape.

Langridge Pistol balls, nails and assorted fragments of hardware fired from cannon for anti-personnel effect. Grapeshot is composed of small iron balls only.

Lias A greyish rock which splits easily into slabs along its lines of stratification.

Limes (plural: Limites) Roman frontier zone.

Machicolation A stone veranda built out from the top of a castle wall. It had holes in the floor for dropping missiles down upon the enemy. These holes were called meurtrières or *murder-holes*.

Maghribi A large Indian catapult weapon.

Manjaniq A huge Indian catapult weapon.

Mantlet Usually a protective covering screening engineers during siege operations; sometimes applied to a detached fortification preventing direct access to a gateway. In the nineteenth and twentieth centuries it was used to describe the armour or other protection screening a gun mounting against direct fire.

Merlon The vertical protective stonework of a battlemented parapet; the opposite of *crenel*.

Migdol Small forts built by the Egyptians.

Moat A ditch full of water.

Mortar-Bound Wall A wall in which the component blocks are set in a pliable substance which dries hard and binds the whole structure solidly together; the opposite of a dry-stone wall in which the component blocks are fitted together without any bonding substance.

Motte The conical mound on which the principal fortification of a Norman castle was built.

Motte-and-Bailey A Norman castle, in which material excavated from a circular ditch was piled up in solidly compacted layers to form a conical mound. The principal fortification was erected upon this *motte*. Either adjacent to or surrounding the *motte* was a large clear area, itself enclosed by a ditch and *rampart*. This was the *bailey*, which served as an exercise area, parade ground and emergency corral for stock. There were also huts for soldiers and skilled artisans.

Multivallate A hillfort with several parallel ditches and ramparts. Hillforts with just two or one such ditch-and-rampart system, are known as bivallate and univallate respectively.

Murder-Holes Holes in castle floors and *machicolation* through which missiles can be dropped upon the enemy; also called meurtrières.

Musculus The Latin word for both a little mouse and a sea-mouse or mussel. It was given to a solid roof or covering (known in later centuries as a *Mantlet*) used during siege operations; in particular one which was leaned against an enemy wall to screen Roman engineers as they tunnelled under the masonry.

Orillons Seen in plan-view, they are the 'barbs' or 'arrowhead' *bastions*. They may be either rounded or acute-angled.

Pa (Pah) Hillfort built by the Maoris in New Zealand.

Below left: The Alcazar, Seville.
Below: The fortified city of Carcassonne.

Palisade A fence of light timbers, usually with small gaps between them.

Parados The protection along the rear of a trench or wall; the exact opposite of a parapet.

Petard A conical cask or other container of gunpowder. The attacker rushed through enemy gunfire to place the wide end against the enemy gate. He then decided upon the best length of fuse to ignite. A correct estimate, and the attacker ran back through the gunfire, while the petard's cone focused the force of the explosion on to the gate and blew it in. Too long a fuse, and the enemy could extinguish it by pouring water down from above. Too short a fuse, and the attacker would literally be hoist with his own petard.

Petraria (plural: Petrariae) A huge stone-throwing catapult of the Roman and Medieval periods.

Picquet A sentry outside an overnight camp. Later it also came to mean the fieldwork for protecting the sentries. During World War I, it was applied to stakes rammed or screwed into the ground for stringing barbed wire. It is sometimes spelt picket.

Pilaster Buttress A buttress is a method of strengthening a wall by building a pillar against it. A pilaster buttress gradually recedes into the structure as it ascends. A flying buttress forms a bridge against the structure.

Portcullis An iron grille, which can be lowered either to keep an enemy out, or to trap him in a *killing-ground.*

Postern A simple doorway for peacetime used by pedestrians, to save opening the main castle gate.

Priest's Cap A freestanding fortification to provide extra fire in front of a Vauban-period fortress; also called a *bonnet.*

Punji Stakes Poisoned pieces of sharpened wood or bamboo employed by the Viet-Cong in Vietnam.

Rampart A raised bank, usually of earth, less commonly of stone. The word is sometimes employed figuratively to describe any sort of defensive wall.

Ravelin A freestanding, near-triangular quadrilateral fortification, located in front of a Vauban-period fortress wall to provide extra fire.

Redoubt A small self-contained fieldwork with earth, stone or sandbag ramparts, and sometimes a ditch beyond. It serves as a base and refuge for soldiers outside the main entrenchments. If a redoubt is actually contiguous to the main fortification, then it may be called a *bastion.*

Re-Entrant Formed when the whole line of the entire wall, entrenchment or other fortified demarcation, retires away from the enemy; the exact opposite of *salient.*

Retaining Wall Built around the base of an earth bank or slope to prevent the dirt spilling forward.

Retirata An improvised fieldwork prepared within that part of a fortress wall which the attackers are trying to breach. Then if they break through, they will still be prevented from rampaging through the whole fortress.

Below: The Citadel at Cairo, founded by Saladin.
Below right: The Spanish fortress of Puerta Real in the Philippines.

Retired Flanks Seen in plan-view, they are the recesses at the base or triangular bastions of the Renaissance period and later. They thus form arrowhead *bastions.*

Revetment To prevent earth banks crumbling, they can be faced and reinforced with wickerwork hurdles, wood panels or dry-stone walling keyed into the soft dirt.

Salient Formed when the whole line of the entire wall, entrenchment or other fortified demarcation, projects towards the enemy; the exact opposite of *re-entrant.*

Sally-Port A small, heavily fortified gateway, from which the defenders can rush out, strike and withdraw hastily before the enemy outside can react.

Sap-and-Parallel A form of siege-warfare introduced by Vauban. Narrow trenches are pushed out in a series of zigzags from a *circumvallation,* towards the fortress. From them lateral trenches are dug parallel to the original encircling entrenchments. The front-line troops then move forward to their new positions. The whole procedure is repeated until close enough for the assault to be launched.

Shell-Keep A circular Norman castle with rooms for accommodation and stores located within the hollow walls.

Stockade A solid fence of heavy timbers; sometimes used for any sort of enclosure.

Switchline A linear fortification or fieldwork, one end resting on the main line of defence. It runs at such an angle that if the enemy breaks through the main defence, then he will naturally swing round to face this new obstacle, inevitably exposing his flank and rear to counter-attack.

Talus A sloping stone apron or plinth, in front of a castle wall or tower; also called a *batter.*

Tenaille A small freestanding bastioned fortification to provide extra fire in front of a Vauban-period fortress.

Testudo (plural: Testudines) The Latin word for tortoise, given to protective coverings (or *mantlets*) screening soldiers during siege operations. The simplest was formed by Roman soldiers locking their shields over their heads and around their bodies.

Trebuchet A Medieval artillery weapon, in which a solid massive weight was attached to the short arm of a pivoted beam. The missile was slung from the end of the long arm, which was then winched down to the ground and released.

Trefoil A three-leaved plant like a clover.

Vallum A Roman earth bank. With a capital letter, it is also the name given to the ditch-and-rampart system marking the southern limit of the military zone along Hadrian's Wall in northern Britain.

Yamajiro A Japanese highland stronghold.

BIBLIOGRAPHY

The subject of military architecture encompasses so many facets of world history that to compile an exhaustive bibliography would be impossible. The selection below, therefore, is a personal recommendation of books on particular aspects of military architecture that the reader may find interesting.

R. Allen Brown, *English Medieval Castles* (B. T. Batsford, London, 1954)

Christopher Duffy, *Siege Warfare* (Routledge and Keegan Paul, London, 1979)

Ian Hogg, *The Guns 1939-45* (Macdonald, London, 1970); plus numerous other highly recommended books on artillery

Anne Johnson, *Roman Forts* (Adam and Charles Black, London, 1983)

Keith Mallory and Arvid Ottar, *The Architecture of Aggression* (The Architectural Press, 1973); describes the twentieth-century fortification of North-West Europe.

Sheila Sutcliffe, *Martello Towers* (David and Charles, Newton Abbot, 1972)

Sidney Toy,
The Castles of Great Britain (William Heinemann, London, 1953)
A History of Fortification (William Heinemann, London, 1955)
The Strongholds of India (William Heinemann, London, 1957)
The Fortified Cities of India (William Heinemann, London, 1965)

Henry Wills, *Pillboxes* (Leo Cooper, London, 1985)

Guide books to particular sites are also an excellent source both of hard facts and interesting anecdotes, and of course encyclopaedias are always worth consulting for information compiled by experts in particular fields. In addition there are numerous magazines and journals covering a variety of military subjects. Of particular interest is *After the Battle*, a series of magazines and other works published by Battle of Britain Prints, London.

The fort at Tallin (Reval) in Estonia.

Index

Page numbers in *italic* refer to the illustrations and captions

Q

R

S

T

U

V

W

Acknowledgements

t. top b. bottom c. centre l. left r. right

p.2, Wales Tourist Board. p.3, J. Allan Cash Photo Library. p.5, J. Allan Cash Photo Library. p.6, M. H. Brice. p.7, M. H. Brice. Back cover, Deal Castle, Kent. p.10, J. Allan Cash Photo Library. p.11, Aerofilms Ltd. p.12, J. Allan Cash Photo Library. p.13, J. Allan Cash Photo Library. p.14, J. Allan Cash Photo Library. p.15, Israel Government Tourist Office. p.16, MARS. p.17, Museum of the Order of St. John. p.18, J. Allan Cash Photo Library. p.20, National Tourist Organisation of Greece. p.23, Museum of the Order of St. John. p.24, M. H. Brice. p.26-7, Israel Government Tourist Office. p.28-9, Mary Evans Picture Library. p.30-31, Mary Evans Picture Library. p. 31, b. J. Allan Cash Photo Library. p.32, Israel Government Tourist Office. p.33, J. Allan Cash Photo Library. p.34, J. Allan Cash Photo Library. p.35, J. Allan Cash Photo Library. p.36-7, Wales Tourist Board. p.38, French Government Tourist Office. p.39, J. Allan Cash Photo Library. p.40, J. Allan Cash Photo Library. p.41, J. Allan Cash Photo Library. p.43, t.l., J. Allan Cash Photo Library. p.43, t.r., A. J. Beckett. p.43, c., Mary Evans Picture Library. p.44, J. Allan Cash Photo Library. p.46-7, Wales Tourist Board. p.47, r., J. Allan Cash Photo Library. p.48, A. J. Beckett. p.49, J. Allan Cash Photo Library. p.50-51 J. Allan Cash Photo Library. p.53, Israel Government Tourist Office. p.55, Museum of the Order of St. John. p.56-7, Museum of the Order of St. John. p.57, r., Wales Tourist Board. p.58, l., Israel Government Tourist Office. p.58-9, GeoScience Features Picture Library. p.60, t., French Government Tourist Office. p.60, b., French Government Tourist Office. p.61, J. Allan Cash Photo Library. p.62, J. Allan Cash Photo Library. p.63, c., J. Allan Cash Photo Library. p.63, b., J. Allan Cash Photo Library. p.64, l., J. Allan Cash Photo Library. p.64, r., Mary Evans Picture Library. p.65, GeoScience Features Picture Library. p.66, J. Allan Cash Photo Library. p.67, Irish Tourist Board. p.68-9, J. Allan Cash Photo Library. p.70-71, French Government Tourist Office. p.72, National Army Museum. p.73, t., J. Allan Cash Photo Library. p.73, b., J. Allan Cash Photo Library. p.74-5, A. J. Beckett. p.75, r., A. J. Beckett. p.76, National Maritime Museum. p.77, Malta National Tourist Office. p.78, t.r., National Maritime Museum. p.78, t.l., National Maritime Museum. p.79, Aerofilms Ltd., p.80, National Maritime Museum. p.81, National Maritime Museum. p.82, Museum of the Order of St. John. p.83, French Government Tourist Office. p.84, National Army Museum. p.85, Museum of the Order of St. John. p.86, Museen der Stadt, Vienna. p.87, Museum of the Order of St. John. p.88, Museen der Stadt, Vienna. p.89, National Maritime Museum. p.90-91, J. Allan Cash Photo Library. p. 92, Museen der Stadt, Vienna. p.93, Museen der Stadt, Vienna. p.94, Mary Evans Picture Library. p.95, t., National Army Museum. p.95, b., National Army Museum. p.96-7, National Maritime Museum. p.97, National Maritime Museum. p.98, Mary Evans Picture Library. p.99, National Maritime Museum. p.100, c.l., National Maritime Museum. p.100, b.r., J. Allan Cash Photo Library. p.103, Japan Information Centre, London. p.104-5, National Maritime Museum. p.106, GeoScience Features Picture Library. p.107, t.l., J. Allan Cash Photo Library. p.107, c., National Army Museum. p.108, Japan National Tourist Organisation. p.110-11, National Maritime Museum. p.111, Japan National Tourist Organisation. p.112-13, National Army Museum. p.115, t.l., National Army Museum. p.115, c.l., National Maritime Museum. p.116, t., Peter Newark's Western Americana, p.116, b., National Maritime Museum. p.117, t.c., North Dakota Parks and Recreation Dept., p.117, m.c., U.S. National Park Service. p.118, Peter Newark's Western Americana. p.119, Singapore Tourist Promotion Board. p.120, British Library, India Office. p.121, Philippines Department of Tourism. p.122, t., Gibraltar Government Tourist Office. p.122, b.l., National Army Museum. p.122, b.r., Gibraltar Government Tourist Office. p.123, Gibraltar Government Tourist Office. p.124, J. Allan Cash Photo Library. p.125, Japan National Tourist Organisation. p.126-7, National Maritime Museum. p.129, National Army Museum. p.130, c.r., National Maritime Museum. p.130, b.r., Imperial War Museum. p.131, American Museum in Britain. p.132-3, National Maritime Museum. p.133, t.r., French Government Tourist Office, p.134, Peter Newark's Western Americana. p.135, c.l., National Maritime Museum. p.135, b., Peter Newark's Western Americana, p.136, t.l., National Army Museum. p.136, c.r., Philippines Department of Tourism. p.137, t.l., Bahamas Tourist Office. p.137, c.r., MARS/U.S. Navy Photo. p.140, J. Allan Cash Photo Library. p.141, National Maritime Museum. p.142, c.l., Imperial War Museum. p.142, b.l., Imperial War Museum. p.142, b.r., Imperial War Museum. p.143, Singapore Tourist Promotion Board. p.144, Singapore Tourist Promotion Board. p.145, Imperial War Museum. p.146, b., French Government Tourist Office. p.147, t.l., Imperial War Museum. p.147, b.l., Imperial War Museum. p.148-9, Imperial War Museum. p. 150, t., Imperial War Museum. p.150-51, Imperial War Museum. p.152, J. Lucas. p.153, MARS/Bundesarchiv. p.155, t.r., MARS. p.155, b., MARS. p.156 Imperial War Museum. p.157, t.r., Imperial War Museum. p.157, b.l., Imperial War Museum. p.158-9, GeoScience Features Picture Library. p.159, b.r., GeoScience Features Picture Library. p.160, t., Imperial War Museum. p.160, b., Imperial War Museum. p.161, t., Imperial War Museum. p.161, b., Imperial War Museum. p.162, t.r., Gibraltar Government Tourist Office. p.162, b., French Government Tourist Office. p.163, 'After The Battle'. p.164-5, 'After The Battle'. p.165, b., MARS/Bundesarchiv. p.166, Defence Staff Information Section, Stockholm. p.167, t.l., Imperial War Museum. p.167, b., Imperial War Museum. p.168, BIPAC. p.169, BIPAC. p.170, Vietnamese Embassy. p.171, MARS. p.172, MARS. p.173, t.l., Vietnamese Embassy. p.173, c.r., Peter Newark's Military Pictures. p.174-5, MARS. p.176, Defence Staff Information Section, Stockholm. p.177, Marinen Informationsavdelningen, Stockholm. p.178, MARS/General Dynamics Corporation. p.179, t., Marinen Informationsavdelningen, Stockholm. p.179, b., MARS/Department of Defense. p.180, Defence Staff Information Section, Stockholm. p.181, t., Modern Alarms Ltd. p.181, b., Modern Alarms Ltd.